Patterns 1
Building blocks

Schools Council Integrated Science Project

The Schools Council Integrated Science Project was set up at Chelsea College, London, from 1969 to 1975. The project team have developed their materials in association with many teachers, and have tested them in a wide range of schools.

Project team

Organisers
W. C. Hall
B. S. Mowl

Team members
J. I. Bausor
Mrs M. P. Jarman
Miss B. A. Lawes
M. R. Nice
D. Wimpenny

Northern Ireland coordinator
S. J. McGuffin

Patterns 1
Building blocks

Authors
William Hall
Brian Mowl

Contributors
John Bausor
David Layton

Published for the Schools Council
by Longman and Penguin Books

Acknowledgements

LONGMAN GROUP LIMITED
London
PENGUIN BOOKS LIMITED
Harmondsworth, Middlesex
for the Schools Council

First published 1973
ISBN 0 582 34009 8

Printed in Great Britain by
Compton Printing Limited
London and Aylesbury

The Authors and publisher are grateful to the following for permission to reproduce copyright material:
Voluntary Committee on Overseas Aid and Development, London, for an extract from *Vox Development Despatch No. 2*; Industrial Press Limited for an extract from *Engineering*; David Layton for Enquiries 5, 6, 7, 10 and 11 of the Carbon section and Enquiry 6 of the Sulphur section from *The Allotropy of Carbon and Sulphur* by David Layton and Times Newspapers Limited for extracts from *The Times*, and *The Times Educational Supplement*.

The authors and publisher are grateful to the following for permission to reproduce photographs: Ardea Photographics, pages 1 (centre), 27 (bottom), 33 (top left and right), 53 (right), 59 and 67 (top left and right); Bexford Ltd., page 123; Professor C. C. Booth, page 93; British Leprosy Relief Association, page 120 (bottom); British Leyland, page 6; the Trustees of the British Museum (Natural History), page 1 (bottom right); Maurice Broomfield, page 83; Burts and Harvey Ltd., page 35 (bottom); Camera Press, page 120 (top); Bruce Coleman, page 12 (bottom) and page 48 (photo: Colin Butler); Commissioner of Police of the Metropolis, page 67 (centre left); Conzett and Huber, Zurich, page 18 (bottom); D. A. Coult, page 71; Philip Damp, page 67 (bottom left and right); Ray Denton, page 21 (top right); Prof. T. Evans, A.E.I., page 98 (top); Electropol Processing Ltd., page 130; Forestry Commission, page 28 (top left) and (top right); Fox Photos Ltd., pages 1 (top left), 1 (bottom left) and 32; Malcolm Fraser and Holt County School for Girls, Wokingham, page 39; Halifax Tool Company, page 21 (top left); Ilford Ltd., page 92; I.C.I. Ltd., page 26 (bottom); Italian State Tourist Office, pages 4 and 21 (bottom left); Keystone Press Agency Ltd., pages 1 (top right), 60 (bottom left) and 67 (centre left); King's College Hospital Renal Unit, page 94; Prof. D. Lacey, page 72; Barry Latter, page 20 (left); Lick Observatory, California, page 15 (top); W. R. Lobb, page 140; Corporation of the Norman Lockyer Observatory of the University of Exeter (photo: W. J. S. Lockyer); London Stone Cleaning and Restoration Co. Ltd., page 5; R. C. McLean and W. R. I. Cook, Textbook of Practical Botany, page 139 (bottom); Mount Everest Foundation, page 57 (left); N.A.S.A., page 16 (top); National Coal Board, pages 27 (top) and 28 (bottom left); National Gallery, page 17 (left); Oxfam, page 64 (bottom); Michael Plomer, page 19; Paul Popper Ltd., pages 1 (top centre), 20 (right) (photo: G. F. Allen), 28 (bottom right), 30, 34 (top), 49 (left), 53 (left), 60 (bottom except left) and 65 (top); The Director, Office of Population Censuses and Surveys, Crown ©, page 42; Radio Times Hulton Picture Library, pages 21 (bottom right), 26 (top left and right), 49 (right) and 119 (bottom); Reed Paper and Board (U.K.) Ltd., page 28 (top centre); Rothamsted Experimental Station, page 139; Sanderson, page 2 (top); Shelter, page 35 (top); Dr James Shields, page 64 (top); Dr D. G. Smith, page 65 (bottom); Syndication International, page 57 (right); Topix, page 2 (bottom); U.K. Atomic Energy Authority, page 124; U.S.I.S., pages 12 (top), 14, 15 (bottom), 16 (bottom), 17 (right) and 98 (bottom); Peter M. Warren, page 141. Cover, Patrick Ward.

Contents

Foreword

This *Pupils' manual* is one of a series designed to provide a scheme of work in science. This scheme is divided into three parts:

Part 1 Building blocks (*Patterns 1* and *2*)
Part 2 Energy (*Patterns 3*)
Part 3 Interactions (*Patterns 4*)

The *Pupils' manuals* are not textbooks containing information for you to memorise. Their purpose is to guide you in experimenting, in examining other scientists' experiments and to help you discuss the ways in which science affects people. The *Pupils' manuals* are only a part of the scheme: you will be given other books to read, films to watch and experiments to perform which are different from, or in addition to, those which appear in these pages.

This is the first of the *Pupils' manuals* and is primarily concerned with building blocks and with the search for patterns, about the way they are made and their behaviour.

A quick glance through the book will show what is meant by the phrase 'building blocks'. The word 'patterns' is also important throughout the entire scheme and so the first section explains its meaning. We hope that you will enjoy searching for patterns, and using them.

1 Patterns and problems

This scheme which you are starting is called *Patterns*. This is because you will be searching for patterns in all that you do in science. The patterns which you find, as well as being interesting, will be useful to you.

Figure 1.1a
Six mammals

What do you understand by the word pattern? What patterns can you see in the drawings and photographs in Figure 1.1? Discuss them with other members of your class. Section 1 will show you the way the word 'patterns' is used in this course. The topic book *The importance of patterns* gives further examples.

Figure 1.1b
Wallpaper pattern

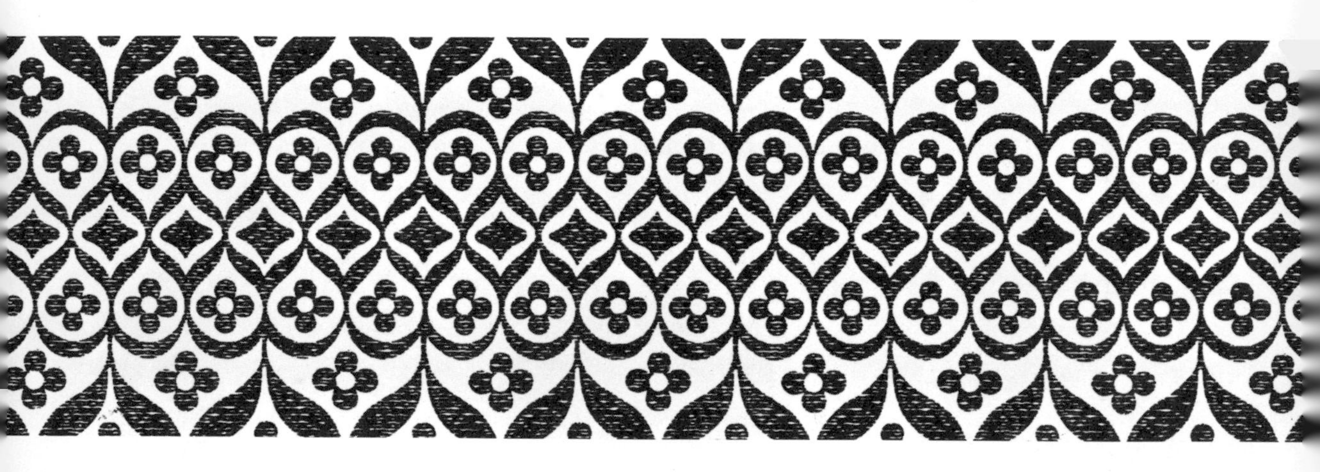

Figure 1.1c
Polystyrene spheres arranged in patterns

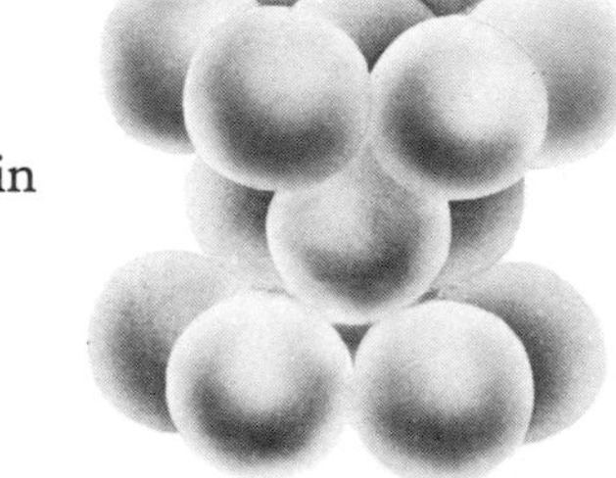

Figure 1.1d

Whenever an important pattern is expected you will see this sign:

It will indicate that there is a generalisation to discuss at that stage of *Patterns*.

▶Problems are shown by black triangles at their beginning and end, as in this sentence.◀ There will also be cartoons to discuss, such as the one shown below:

Is it just 'opinions' we are considering in science or are patterns rather more than this?

▷ This sign indicates optional work.

Investigation 1.1 The pattern of acids and carbonates

You will need

test tube, 125 × 16 mm
spatula
stirring rod
lime water
dilute hydrochloric and nitric acids
copper carbonate, zinc carbonate, nickel carbonate and cobalt carbonate

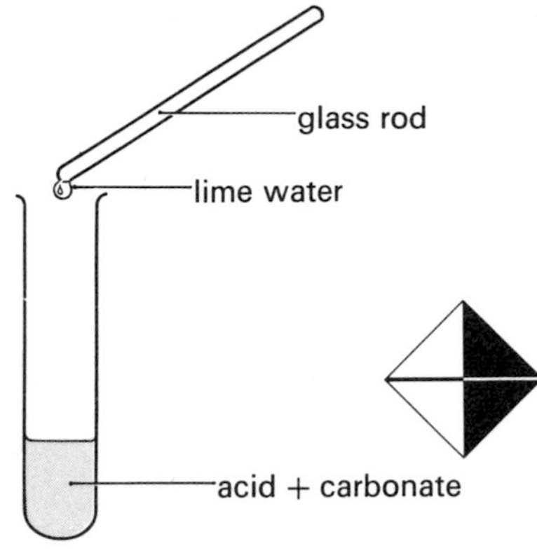

Figure 1.2

Put one measure of copper carbonate into the test tube. (Here, and throughout this book, a measure means the amount of substance obtained by using the Nuffield spatula.) One-quarter fill the test tube with dilute hydrochloric acid. Test any gas which is evolved with one drop of lime water; figure 1.2 shows one way of doing this. If the lime water turns 'milky', the gas evolved is carbon dioxide. Repeat the experiment for the other carbonates, and then use dilute nitric acid instead of hydrochloric acid.

Write a sentence about the effect of acid on carbonates.

This is a pattern.

Patterns are useful to us because they can be employed to solve problems. Here are three problems for you to solve using the pattern you have just discovered.

Investigation 1.2 Making a prediction

▶Predict the action of dilute sulphuric acid on copper carbonate and then test your prediction.◀

Investigation 1.3 Hunting the baking powder

▶Labels have fallen off jars of cornflour, French chalk, icing sugar and baking powder. Baking powder contains a carbonate, and there is a bottle of vinegar available (vinegar is an acid). Identify the baking powder.◀

Investigation 1.4 The problem of marble

Use dilute hydrochloric acid to show that marble is a carbonate. ▶Predict the action of dilute sulphuric acid on marble. Test your prediction experimentally. Give an explanation for your observation.◀

Now modify the pattern to take account of the observation in Investigation 1.4.

In Investigation 1.2 you were able to make a prediction from the pattern. An everyday problem could be solved in Investigation 1.3 and the pattern had to be modified after Investigation 1.4.

Figure 1.3
Blocks of marble. Are there any buildings faced with marble in your locality?

In the next investigation you will be discussing one of the everyday examples of the pattern; another example (hardness of water) is given in *The importance of patterns*.

Investigation 1.5 Weathering

Figure 1.4
Effects of weathering

Figure 1.4 shows the effect of 'weathering' on old limestone buildings. What has been attacked? Explain one of the ways in which this might have happened. What problems arise because of erosion such as this? Look for similar examples in the area of your school.

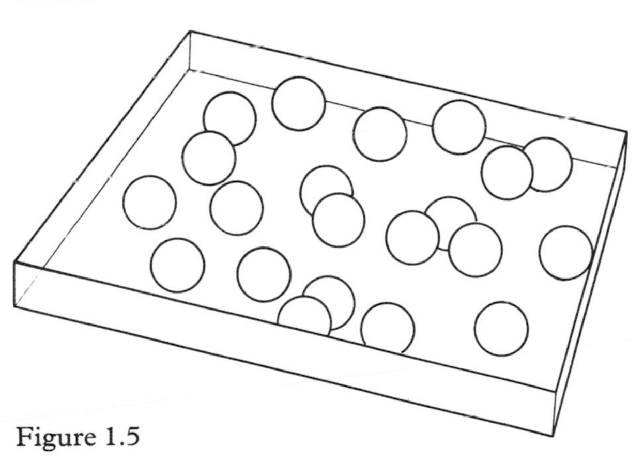

Figure 1.5

In discovering the previous pattern you have experimented with solids, liquids and a gas. Let us take a closer look at solids, liquids and gases. How do they behave when they are put into a container? In the next investigation you will try to explain these patterns using a model which is shown in figure 1.5.

Investigation 1.6 Marbles as models

You will need

shallow tray
wooden partition
forty marbles, about 1 cm diameter

Keeping the tray flat on the bench, move it in an irregular manner. What are the patterns of movement of the marbles?

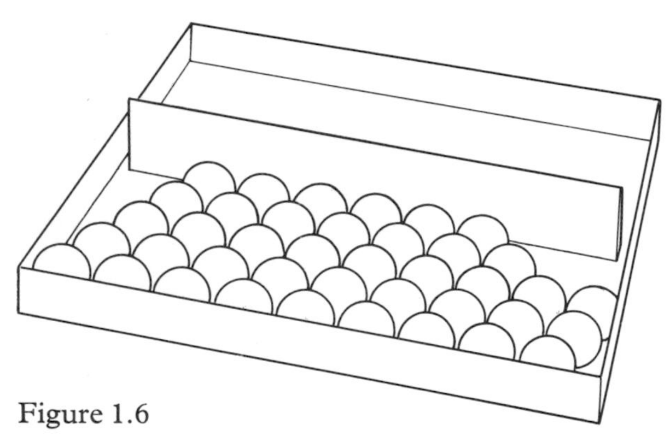

Figure 1.6

Repeat the experiment, but this time with the wooden partition in place, as shown in figure 1.6. How are the patterns of movement modified? Let us assume that gases consist of tiny particles dashing around but that the particles in liquids and solids are closer together. How does the model help in explaining the properties of gases, liquids or solids which you listed at the start of the investigation? In what ways is the model not helpful?

Figure 1.7a
The braking system of this car is controlled hydraulically

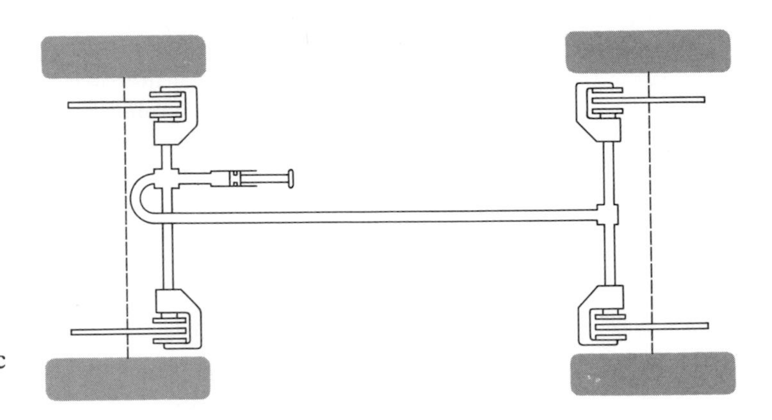

Figure 1.7b
What properties of liquids are particularly useful in hydraulic brakes?

Investigation 1.7 Predicting pushes

You will need

a syringe

▶Use the previous model to predict what would happen when you try to compress a solid, a liquid and a gas. Devise experiments to test your prediction.◀ This model will be used on future occasions to explain properties of solids, liquids and gases. Remember: it is only a model; particles in solids, liquids and gases are not round and hard like marbles!

▷Investigation 1.8 Comparing solid materials

You will need

you must decide what apparatus is needed

You are provided with a collection of solid objects made of several different materials. Complete a table like that shown for as many of

object	material	colour	surface texture	mass/g	volume/cm^3	hardness

the objects as possible. You may need to devise suitable ways to compare or measure the properties of the objects. Which of the properties in your table are useful in identifying an unknown material? What other properties could be used for this purpose?

Nearly all such properties depend on either the appearance of the substance, or the results of testing it in some way. Is there any property of a substance which could enable you to identify a piece of it if:

a it was painted or coated in some way

b you were not allowed to damage the coating?

The following suggestion should enable you to answer that question.

Plot a graph of the entries in columns 5 and 6 of your table as shown, using a different symbol for each material. What conclusion

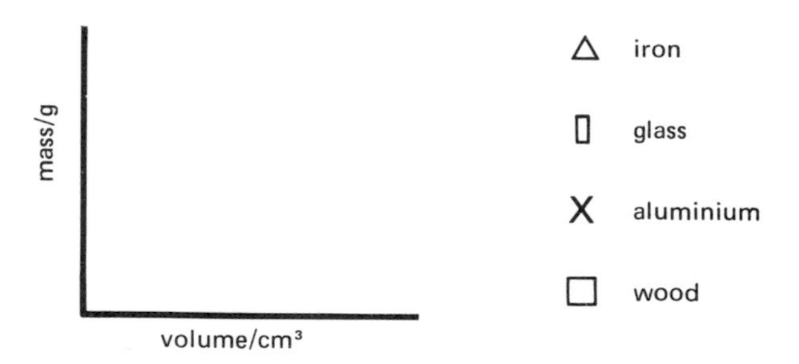

can you draw from your graph? Use the graph to formulate a simple mathematical pattern about the relation between mass and volume.

This relation illustrated by your graph (for a particular substance) is one called 'proportionality'. We say that the mass of a piece of iron is proportional to its volume. (We can equally well say its volume is proportional to its mass.) For different objects of the same material, the mass and the volume can be different. Is there any related property which is the same for all these objects? The graph (and also the table) should enable you to find one (if necessary add an extra column to your table). Does finding a property of this sort depend on the relation of proportionality or would there still be such a property if the graph were different?

Use your graph and the pattern of proportionality to find the density in $g\ cm^{-3}$ of:

a iron

b glass

c water.

A slide-rule is useful for this sort of calculation. What is the density of each of these materials in $kg\ m^{-3}$?

Use your graph and the pattern of proportionality to identify the material of which a painted object or block is made. Is there any doubt about the identification? What other statement can be made about the material without doubts?

Are your measurements of mass and volume exact? If not, which is the more precise? What does this imply for the relation between them? What does it imply about the density?

The relation of proportionality is a very important one in science. Why do you think this is? Try to suggest other pairs of measurable quantities which you know are related in this way. What other pairs are there which might be related in this way, but which you can only tell by actually doing experiments or making measurements?

The next investigation is of this sort. Some solids can easily be stretched, and rubber is one such example; you will be searching for a pattern in the stretching of rubber.

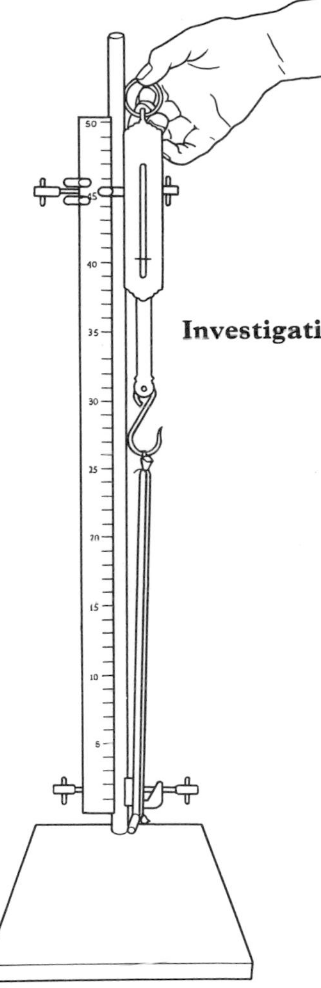

Figure 1.8

Investigation 1.9 Pulling rubber

You will need

metre rule
rubber cord, 55 cm
identical rubber bands
two pieces of string
retort stand, two clamps and bosses
force meter

Part a

Do a rough experiment with a piece of rubber cord or a rubber band to see about how much it can be stretched, and use a force meter to find the size of the force needed to stretch it this much (see figure 1.8).

Some people can now do part b and some part c.

Part b

Devise an experiment to find the relation (pattern) between the amount the rubber stretches (the extension) and the force stretching it. Use tabulated results and a graph as shown on page 10 to represent the pattern. You can also use a mapping diagram if this is familiar to you.

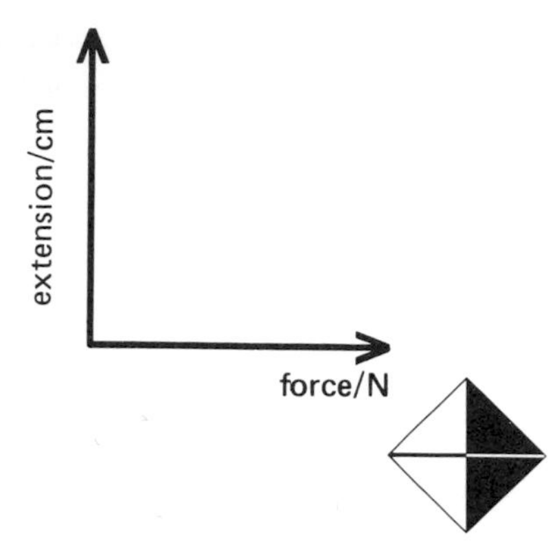

length of rubber/cm	extension (stretch)/cm	force (pull)/N
20	0	0
30	10	
40	20	
50	30	

Part c

Devise an experiment to find the relation (pattern) between the amount a piece of rubber stretches (the extension) and the unstretched length for a fixed size of force. (You could choose a force which stretches the shortest length of rubber by half its length, or a force which doubles the length, but you should continue to use whatever force you use first.) Use tabulated results and a graph as shown below to represent the pattern. You can also use a mapping diagram if this is familiar to you.

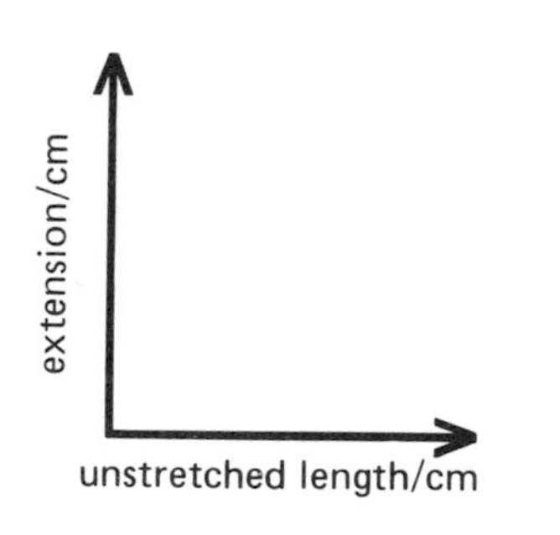

unstretched length/cm	stretched length/cm	extension/cm
10		
20		
30		
40		

In each case (parts b and c) is the pattern one of proportionality? If so, what corresponds to density in the previous investigation? If not, describe how the pattern differs from the pattern of proportionality.

Could either of these patterns be predicted from what you knew before doing the experiment?

▷You could do additional experiments to find out whether other materials behave in the same way as rubber.

Investigation 1.10 Predicting pulls

You will need

rubber bands (identical)
you decide what else you will need and how to arrange the apparatus!

▶Find the force necessary to stretch one rubber band by one-quarter of its length.◀

▶Predict the force necessary to stretch two rubber bands (side by side) by one-quarter of their length. Now make the same prediction for three, four and five rubber bands.◀

▶Test your predictions experimentally.◀

▶You might like to make a measuring instrument at home for measuring pulls. Try to devise one for measuring pushes. It could be calibrated at school.◀

Investigation 1.11 Proportionality again

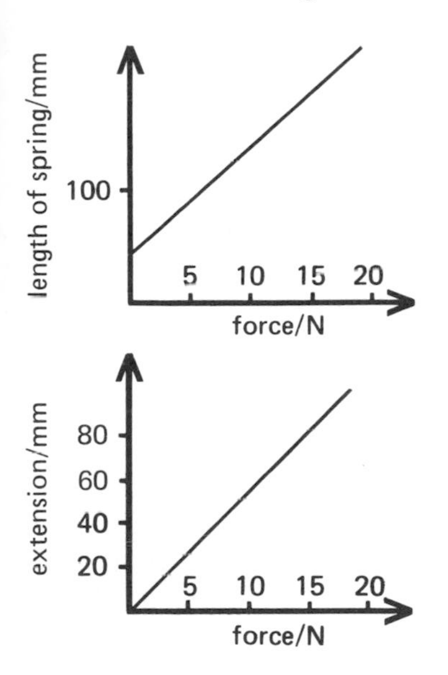

The following graphs show the results of an experiment using a steel spring, represented in two different ways. You may have done such an experiment yourself.

Notice the different scales on the vertical axes. ▶Which relation (pattern) is one of proportionality? What would be a suitable description of the other relation? Find an equation for each relation. Which equation is simpler? Use your answers to explain why it was convenient in the experiment pulling rubber to calculate the extension and use this in plotting a graph. How is it possible to convert a linear relation into a relation of proportionality?◀

In this section you have been discovering patterns and using the patterns to make predictions and to solve other problems. You will be doing this throughout the three years. The patterns will become more complex and the problems will get more difficult (and more numerous!) as the course progresses.

Figure 1.9
What interesting patterns there must be in order to launch a command module or to understand animal behaviour

2 Galaxies, planets and the Earth

Would you like to live on another planet? What are the special features of the Earth which make it possible for us to live here? Already you must have opinions on questions like these. Space flight and satellite photographs have given us much information about our solar system and at the beginning of this section you will be looking at some of these data.

Investigation 2.1 The Earth, a planet of life

Figures 2.1a – c on pages 14–15 show photographs of the Earth, the Moon and Mars. They are three of the building blocks within our solar system. Compare the diameters of the three planets. Methods of obtaining these diameters are discussed in *Length and its measurement*.

The turbulent cloud patterns over the face of the Earth are due to movements in the atmosphere. Can you detect the existence of an atmosphere on either the Moon or Mars?

What other distinctive features can you see on these pictures taken from a very great distance? You could summarise your observations in the form of a chart like the one following:

	Earth	Moon	Mars
diameter/km			
atmosphere (yes/no)			
(add your own headings)			

Figure 2.1a
The Earth

5 000 km

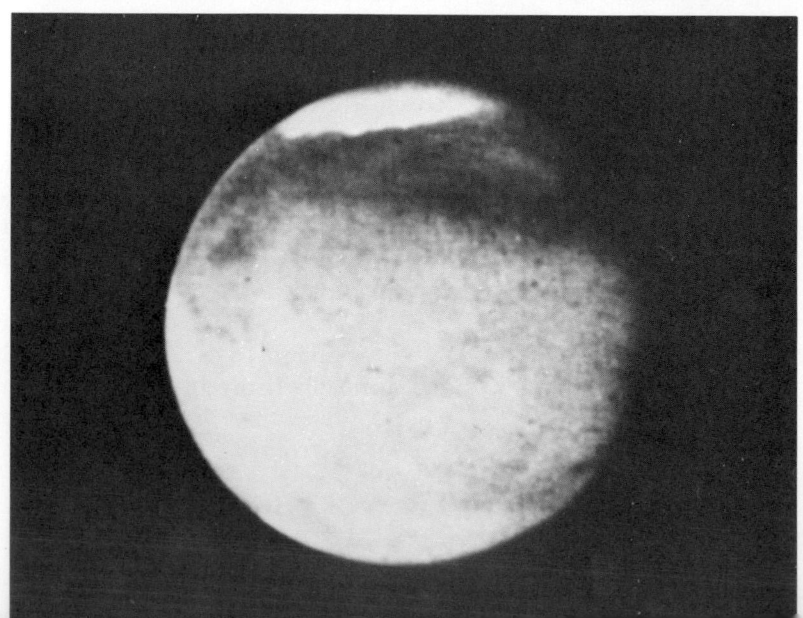

Figure 2.1b
Mars

3 000 km

Figure 2.1c
The Moon

1 000 km

Now compare the three closer views shown in figures 2.2a – c. Note the variations in the scales of these photographs, and the difference between the vertical and oblique shots. Which two have similar relief features? Describe the shapes and sizes of these features. Suggest ways in which they may have been formed.

What sort of features can you recognise in the photograph of the Earth? Which of these features can be related to the fact that the Earth has an atmosphere? Can any of the features be related to the fact that there is life on the Earth?

Figure 2.2a
A closer view of part of the Earth

100 km

Figure 2.2b
A closer view of part of Mars

100 km

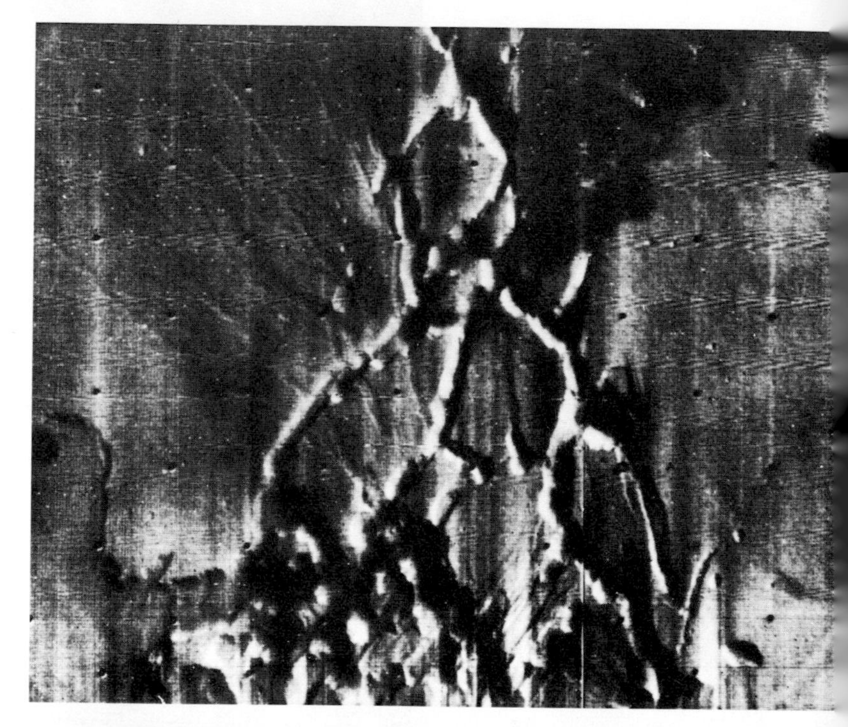

Figure 2.2c
A closer view of part of the Moon

15 km

Figure 2.3a is part of a painting by the eighteenth century artist Vernet. Figure 2.3b shows Apollo 11 astronaut Edwin Aldrin walking on the surface of the Moon.

The astronaut is carrying scientific equipment and this makes him seem unduly cluttered. Why does his clothing contrast with what he would wear on the Earth? What is the nature of the Moon's surface? Describe the general view and the small-scale features.

How does this detailed picture of the Moon's surface contrast with the picture of the Earth's surface? Are there parts of the Earth's surface which resemble the surface of the Moon?

Figure 2.3a 'A seashore' by Vernet

Figure 2.3b Edwin Aldrin on the Moon's surface

Some of the American astronauts claimed that their flights provided evidence of God's existence, and yet the Russians claimed that the reverse was true. What do you think?

Investigation 2.2 Planets in motion

Figure 2.4a

Figure 2.4b

Figure 2.4a was obtained by pointing a camera at the Pole Star and leaving the shutter open for about two hours. ▶ Does the photograph suggest to you that the stars are in motion or are stationary? Are there other possibilities? ◀ The photographs of figure 2.5a – f will help in your discussions.

(a) Subject, background and camera all still

(b) Subject moving, background still

(c) Subject still, background moving upwards

(d) Was the camera moving?

(e) Who was moving sideways?

(f) What was moving sideways?

Figure 2.5a–f

Figure 2.4b shows the complete path of the sun as seen at mid-summer in the Arctic.

Imagine the photographs to be arranged to form a cylinder. ► How does the sun seem to move? Is there another explanation? ◄

You will be reading more about planetary motion later in the scheme. Planets are the building blocks of the solar system, and the solar system is part of a much bigger system known as a galaxy. The planets themselves consist of smaller building blocks which include atoms, molecules and organisms. In the rest of this section you will be taking a brief look at some of these building blocks and the way man uses them.

The solid crust of the Earth is made up of many different rock types. Think of some of the different rocks you have seen. How many of these rocks have you examined closely? The following investigation will enable you to see how the diversity of rock types making up the crust can be grouped into three major categories.

Investigation 2.3 Looking at rocks

You will need

one granite specimen
one basalt specimen
one sandstone specimen
one limestone specimen
one marble specimen
one slate specimen

Figure 2.6a – f
The photographs show possible locations for the six specimens

(a) granite

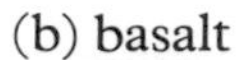

(b) basalt

(c) sandstone

(d) limestone

(e) marble

(f) slate

Examine each specimen in turn and try to answer the following questions, the book *Rocks and minerals* should be used in this:

1 Are the components visible as individual particles?
2 If so, what shapes do the particles have?
3 If components are not visible, is the rock glassy in texture?
4 Is there a layered structure visible –
5 and/or a general banded structure?

Assign each of the specimens in turn to one of the major groups and comment briefly on their origins.

Both living and non-living things can be of use to man. For example, we rely on living things for food and for clothing; marble and granite are used in buildings. The next investigations illustrate how common non-living materials can be employed to obtain materials which are useful to man. You will find further examples in the Nuffield Chemistry background book *Chemicals from Nature* and the Longman Physics Topics book *Materials*. The last part of this section gives examples of how living materials are used by man.

Investigation 2.4 Getting copper from its oxide

Some elements burn more brightly than others in oxygen. It is possible for one element with a strong 'attraction' for oxygen to remove the oxygen from the oxide of another element. Your teacher will demonstrate how copper can be obtained from its oxide by using magnesium.

Investigation 2.5 Extracting iron from iron ore

The film loop *Iron Extraction* and the wall charts will show how iron is extracted industrially. In what ways are the reactions similar to those in Investigation 2.4? Note the scale of the industrial process. What is iron used for?

Salt is one of the most common chemicals (yes! it is a chemical) that you will ever meet. Apart from its obvious uses (what are they?) salt may be used to obtain another important chemical. This can be achieved electrically.

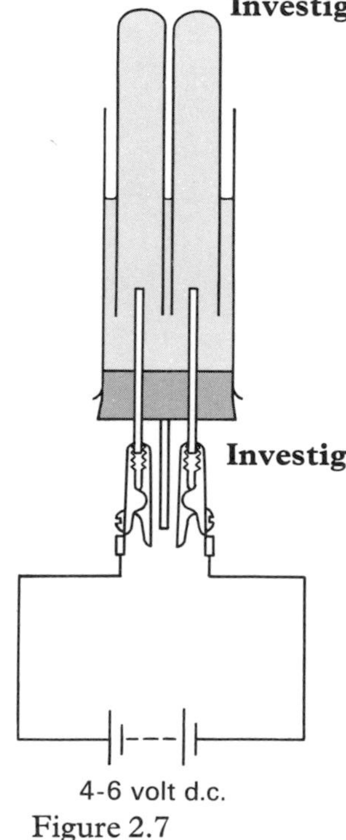

Figure 2.7
Electrolysis cell

Investigation 2.6 Obtaining a useful gas from salt

You will need

electrolysis cell as illustrated in Figure 2.7, with two rimless test tubes
two lengths of connecting wire with crocodile clips
six-volt battery or d.c. supply
splints
stand and clamp
universal indicator paper
salt solution or sea-water

If you have previously electrolysed sea-water, you could try copper sulphate, or copper chloride, or silver nitrate instead.

Electrolyse the sea-water in the apparatus. Try to find out what is evolved

a at the positive electrode (the anode)

b at the negative electrode (the cathode)

by testing the products with moist indicator paper and a lighted splint. Starch/iodide paper could also be used. The following chart, which summarises tests for various gases, will be useful here and later in the scheme.

Gas	Litmus	Other tests
oxygen	no change	rekindles a glowing spill
nitrogen	no change	no positive test (spill extinguished)
hydrogen	no change	causes a 'pop' when lighted spill applied to air and hydrogen mixture
carbon dioxide	blue → slightly red	lime water goes milky
carbon monoxide	no change	burns with blue flame
chlorine (green)	blue → red → white	starch/iodide paper goes blue
bromine (brown)	blue → red	starch/iodide paper goes blue
iodine (violet)	blue → red	
hydrogen chloride	blue → red	misty-white fumes with ammonia
sulphur dioxide	blue → red	acidified potassium dichromate paper goes green
hydrogen sulphide	blue → red	lead nitrate paper goes black
ammonia	red → blue	milky-white fumes with hydrogen chloride

Smell the products and the electrolysed solution.

The film loops *Chlorine Manufacture* and *Chlorine Uses* will show the industrial manufacture and uses of one of the gases you obtained.

Investigation 2.7 Lead from its oxide

▶Devise a method of obtaining lead from its red oxide and discuss it with your teacher. Is energy needed to do this?◀ What is lead used for? Was energy required in each of the extraction processes? Could there be a pattern here? Carry out the experiment.

▷Investigation 2.8 The scale and economics of some industrial processes

So far you have carried out small-scale experiments. The films *Filtration, Crystallisation* and *Distillation* give some idea of the scale of industrial processes. In this investigation you are concerned both with the scale and with the economics of one process.

You will need

wall charts and booklets dealing with any one of these industrial processes
iron ore to steel
Solvay process
manufacture of an acid
making a plastic

Using reference books, write details of each chemical reaction in the process. (No equations are needed unless you understand what you are writing.) Draw a flow chart to help in this.

List the expenditure required to make the product under two headings:

a Capital expense
b Recurring expense

Under capital expense include buildings, machinery, etc., and

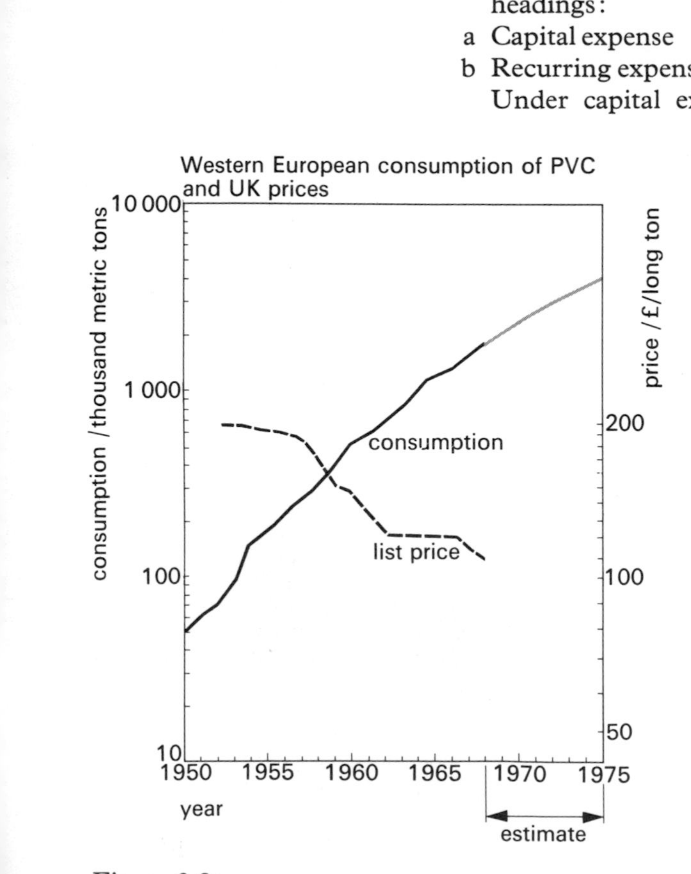

Figure 2.8a

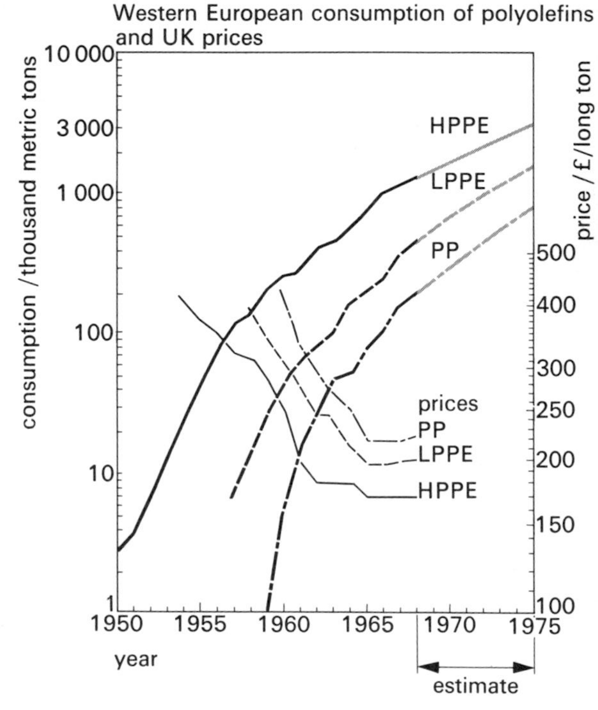

Figure 2.8b

under recurring expense list fuels, labour, raw materials, etc.

Suggest those factors which could cause your final product to decrease in price after a number of years and those which could cause a price increase.

Who would buy your product? Would much land be required to build the factory?

The graphs (figures 2.8a and b) show how the prices of three types of polypropylene plastic have altered when production has increased. How has cost been affected by production?

How are large quantities of chemical handled, and what are some of the problems of carrying out reactions on this scale?

What industries are situated in your locality? Are the buildings ugly? Does this matter? Do the industries provide much employment? Is this more important?

How metals fared in 1971

Although it was a generally bearish year for prices, turnover on the London Metal Exchange last year reached record levels in copper, lead and zinc. There has been an unbroken increase in sales of copper and lead since the mid-1960s. Zinc was the only metal to show an increase in price over the year and this is reflected in turnover. Silver turnover disappointed again and it has fallen each year since the 1969 record when sales were 379 830 000 ounces.

Metal	Turnover tonne 1971	Turnover tonne 1970	Prices* 1971 High	1971 Low	Prices* 1970 High	1970 Low
Copper	2 888 000	2 670 950	£535.75	£393.75	£748.50	£421
Tin	144 850	151 970	£1499	£1399.50	£1637	£1433
Lead	778 700	709 875	£112.75	£85.125	£144.875	£109.125
Zinc	640 225	296 775	£144.625	£111.875	£127.875	£118.125
Silver†	309 490 000	330 620 000	72.3p	51.2p	80p	65.5p

*Middle closing cash quotations in afternoon ring trading. †In troy ounces and pence.

Discuss the figures shown in the table above. Which metals have a decreased turnover from 1970–71? How do 1970 prices compare with those in 1971?

Investigation 2.9 Distribution of minerals

Consult reference books to find where different ores are distributed throughout the world.

Which countries have supplies of the following?

a copper
b uranium
c gold.

Which minerals are mined in the UK? Which parts of the world seem to be richest in mineral deposits? Are the parts of the world richest in minerals also richest in terms of their people's prosperity?

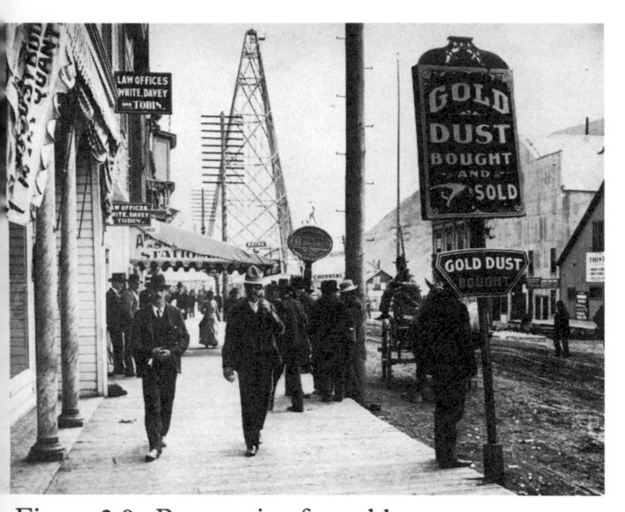

Figure 2.9a Prospecting for gold

Figure 2.9b A mining village

How can the presence of minerals affect the movement of people in a country? What happens to those communities when the mineral deposits are used up?

Investigation 2.10 The effects on the countryside of man's use of materials

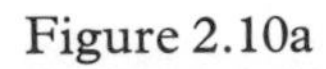

Figure 2.10a

Figure 2.10b

The cheapest method of obtaining materials can sometimes destroy the natural beauty of the countryside. The photographs of figure 2.10a and b show two views of ICI's Tunstead Quarry. Write an account of your reactions to what is happening there. It should be borne in mind in your account that we do need rocks and minerals!

Attempts are also made by industry to reclaim land, as figure 2.10c shows. Is this a possible solution to the problem?

Figure 2.10c

The Earth supports an incredible variety of living things. Ponds like that shown in figure 2.11 may be an everyday sight, but life teems around them, in them and even on them. How many different plants are there growing around the pond in the photograph? Is the pond similar to any you know? If so, have a close look at the life associated with it. Do you think it is important to retain this life? Why do living things sometimes become extinct?

Figure 2.11

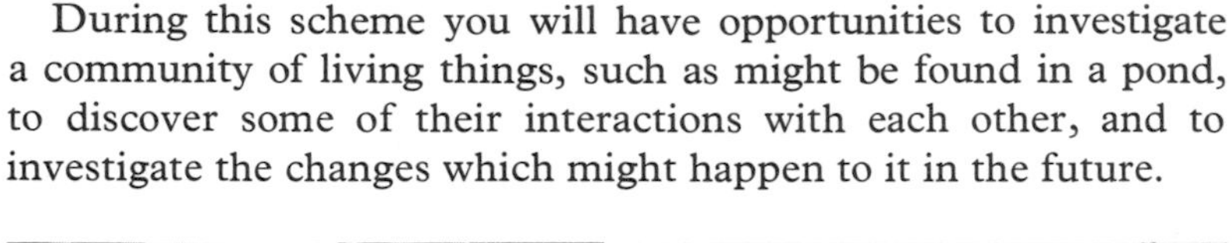

During this scheme you will have opportunities to investigate a community of living things, such as might be found in a pond, to discover some of their interactions with each other, and to investigate the changes which might happen to it in the future.

Figure 2.12

(a) Wood is one of the more widely used materials from living organisms

(b) Pulping wood for paper production

(c) Continuous replanting must take place

(d) Planting young trees on coal tips

(e) Wood is still one of the most beautiful construction materials for furniture

(f) Wood cladding on a modern house

Of all the materials useful to man which are obtained from living things wood is probably still the most common. In the past it was even more widely used but it has now been partly replaced by other materials. What were some of these former uses and what has replaced the wood? Why do you think wood is still used for many purposes such as in buildings or for furniture?

Which materials have been used to make the clothes you are wearing, the food you eat, the furniture you use and so on?

Already (Investigations 2.5–2.8) you have obtained important substances from non-living materials. In the next experiment you will be obtaining iodine from a living material – seaweed.

Investigation 2.11 Iodine from seaweed

You will need

beaker, 100 cm^3
test tube, 150 × 25 mm
Bunsen burner, tripod, gauze and hardboard mat
filter funnel and filter paper
dried seaweed *(Laminaria)*, about 1 g
twenty-volume hydrogen peroxide
sulphuric acid
tetrachloromethane
distilled water

Gently boil the seaweed with 10 cm^3 distilled water in a beaker for a few minutes. Filter the solution and collect the filtrate in the test tube. Add 2 cm^3 of sulphuric acid and 10 cm^3 of hydrogen peroxide to the filtrate in the test tube. Now add 3 cm^3 of tetrachloromethane and shake. If iodine is present, the tetrachloromethane should turn purple. By pooling all of the purple solution of iodine in tetrachloromethane which has been obtained by the class, your teacher can obtain crystals of iodine.

The film loops show how iodine is manufactured on a large scale and also shows some of its uses.

The Earth is a planet of life containing a large variety of living and non-living building blocks. The rest of *Patterns 1* is concerned with a study of some of these building blocks. Their connections are illustrated below:

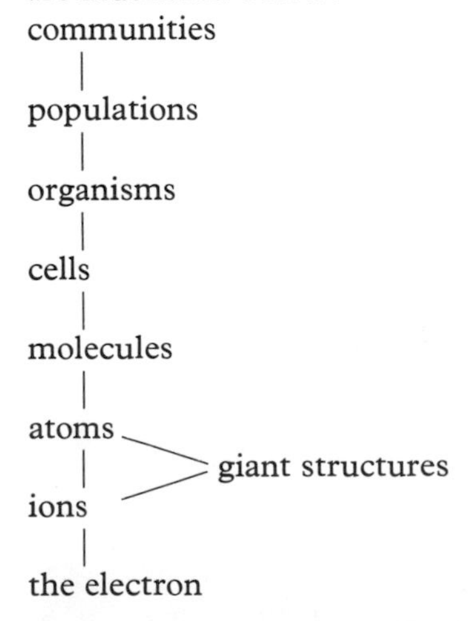

3 Communities and populations

Figure 3.1
A tropical forest. The photograph shows a variety of plants, but what animals could be found living in a tropical forest?

> The forest at night was a very different place from the forest by day: everything seemed awake and watchful, and eyes gleamed in the tree-tops above you. Rustles and squeaks came from the undergrowth, and by the light of the torch you could see a creeper swaying and twitching, indication of some movement one hundred and fifty feet above you in the black tree-top. Ripe fruit would patter down on to the forest floor, and dead twigs would fall. The cicadas, who never seemed to sleep, would be screeching away, and occasionally a big bird would start a loud 'car . . . carr . . . carr' cry, which would echo through the forest.

From 'The Drunken Forest' by Gerald Durrell

One of the most striking differences between the Earth and the Moon is the presence and absence of living organisms – plants and animals. Indeed, the Earth might be imagined as being covered almost entirely with a blanket of life on the land, in the air and sea. The Earth's 'green mantle' is one way it has been described. In his book Gerald Durrell gives his impression of a tropical forest in the Cameroons in West Africa, just a small part of the Earth's green mantle. Any forest or woodland you cared to investigate would be found to consist of a bewildering variety of organisms, seen and unseen, living together.

How do organisms living together affect each other? How do they differ? How are they distributed? These are just some of the questions you will consider in your search for patterns in this course.

Investigation 3.1 The Earth's green mantle – a pattern of distribution

Study the map (figure 3.2) showing the distribution of the main types of vegetation covering the Earth.

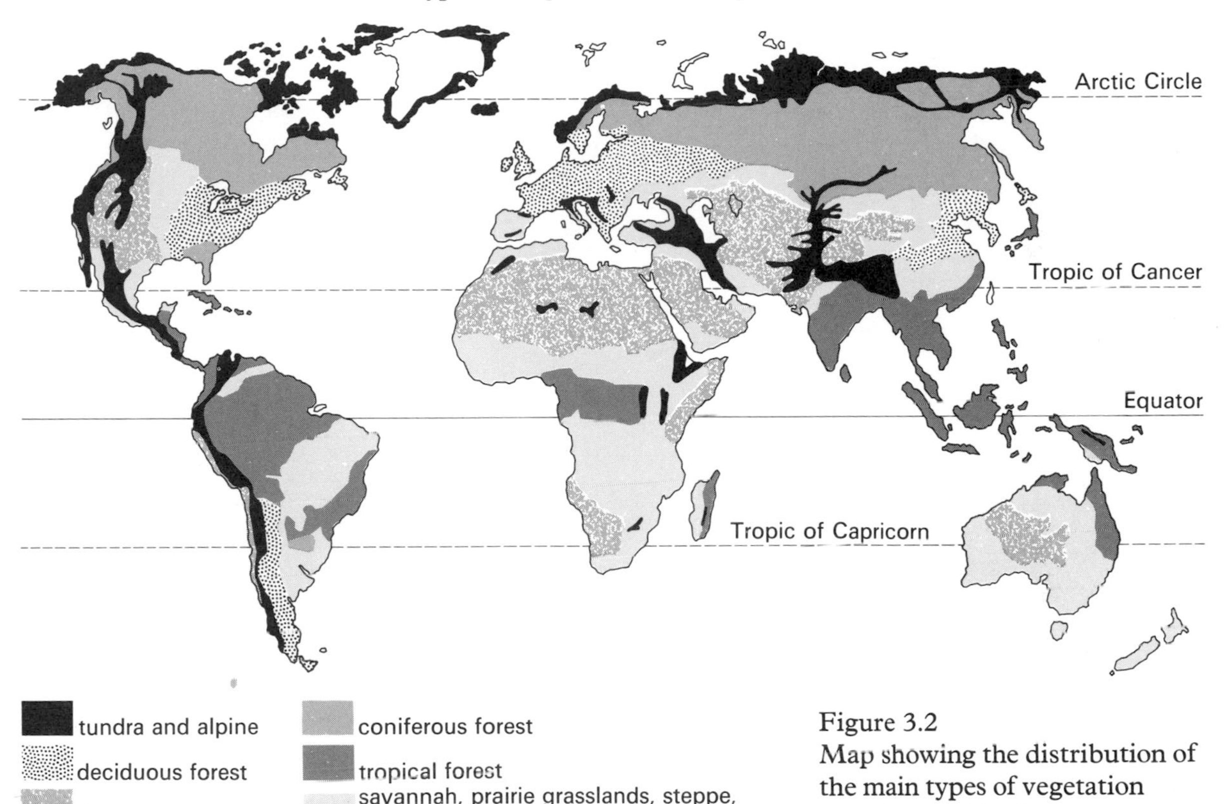

Figure 3.2
Map showing the distribution of the main types of vegetation covering the Earth

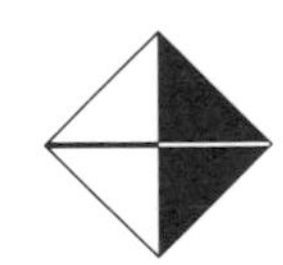

What is the pattern of distribution of the vegetational types? Suggest an explanation for this pattern of distribution. Use other sources, such as an atlas, to find out if your explanation is correct. ►How would you account for the distribution of tundra and alpine vegetation down the Pacific coastline of the Americas?◄

Any system of different organisms living together such as a forest, a woodland, or a pond, is called a community. One way of regarding the major areas of vegetation on the Earth is as very large community systems each with their distinctive variety of organisms. It would be possible to regard the entire Earth as an even larger community system. On the other hand, communities can be much smaller. It depends very much on where you, as an

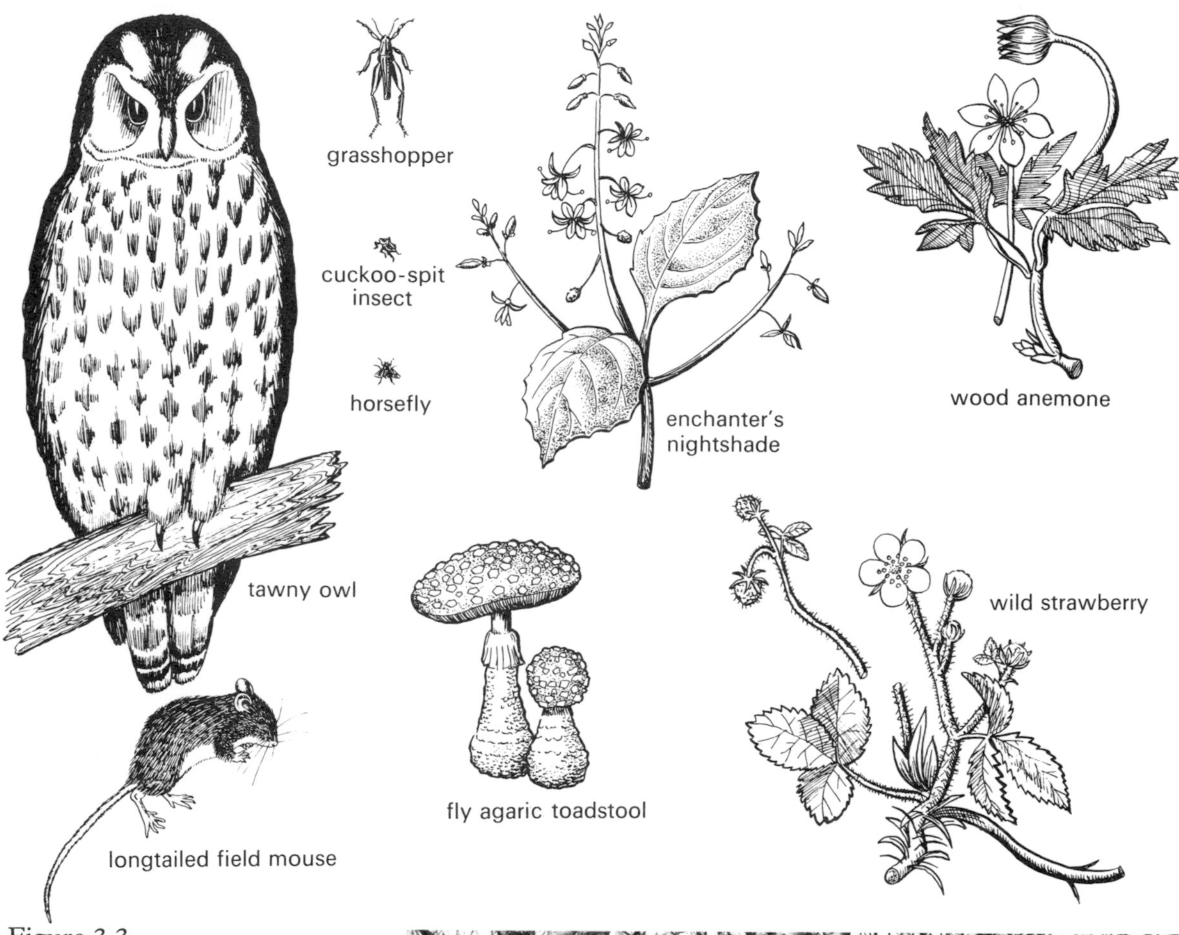

Figure 3.3
The variety of life in a typical deciduous woodland such as is commonly found in Britain. Deciduous trees are those which lose their leaves in winter

Figure 3.4 Typical savannah grassland

Figure 3.5 Tundra

individual, choose to draw an imaginary boundary round a collection of different organisms living together. It might suit one person to regard a tree, with all its associated plants and animals, as a community system for investigation. It might suit another person to regard a whole wood as a community system.

Investigation 3.2 Looking at communities

You will need

warm clothes, if it is cold
pencil and notebook

Look for community systems around your school. Later, look for others on your way home.

Construct a map (roughly to scale) showing some of the community systems around the school (see figure 3.6), giving each a suitable name: lawn community; hedge community and so on. Do some overlap? Are any distinct, but 'inside' a larger community system? If possible name the organisms making up the community (but this is not essential). Do any organisms in the community obviously affect, or interact with, others?

Figure 3.6
Two examples of community systems which you might easily find around your school

Figure 3.7
How many community systems?

Make a note of the main features of the surroundings, the environment, which you suspect may affect, or interact with, each community you map. For example, it might be sunny or shady, it might be dry or wet, it might face north or south and so on. A community system, together with the features of the environment which may affect it, is called an ecosystem.

Compare the map you have drawn with others. Is there a common pattern to the community systems you have identified and mapped? How much does man affect the communities studied?

Investigation 3.3 The school community

Figure 3.8

A community system can be regarded as being made up of a variety of different organisms living together, but we often talk about groups of human beings as being communities. The people who work in your school make up a community. How does this idea of a community system differ from that used in Investigation 3.2? Are there any similarities?

Does the school community fit into a much wider community system? Discuss the human community systems in your area. Many of your lessons in school may be concerned with studying human community systems and how the different parts affect each other.

Describe your reactions to the contrasting human communities shown in figures 3.9 and 3.10.

In Investigation 3.2 you were asked to decide how much man may affect the communities you studied. One of the great concerns of our time is the way in which man affects his surroundings which include communities of the sort you have been studying. Nearly everyday in newspapers it is possible to read about the effects of pollution for example, or maybe there is an outcry because the local council wishes to cut down some trees to widen a road.

Figure 3.9b

Figure 3.9b

Figure 3.9c

Figure 3.10a

Figure 3.10b

Figure 3.10c

To answer the question posed by figure 3.11 it is necessary to try to find out patterns in community systems in order to be able to understand, judge and predict the effects of man's activities. Since the main effect of human action will probably be to cause some sort of change the first step is to find out what patterns of change, if any, take place when man does not obviously interfere with a community system.

Figure 3.11
Spraying a roadside verge with weedkiller. Is it possible that this will cause changes to the community other than killing the weeds?

Investigation 3.4 Patterns of change in a community

You will need

three troughs, preferably plastic, one for each group
plentiful supply of filtered pond water
leaf mould
dried mud from the edge of a pond
glass-writing pencil

As you will have realised from earlier work, many community systems are complicated. They may consist of a very large number of different organisms or they may merge into other community systems. In order to overcome some of these difficulties you can make use of a mini-pond community in an aquarium.

Cover the bottom of the aquarium with some of the dried mud, add a little leaf mould and fill with pond water. Place the aquarium outdoors on a roof or in a position where large animals (particularly dogs and young humans) cannot interfere with it. Mark the initial water level with a glass-writing pencil on the side of each vessel and, when necessary, make up the water lost through evaporation. This is as much as you need to do now. At some suitable time in the future (you will be told) you must make your first examination of the aquaria for organisms. Instructions are given later in this section. By continuing the investigation in the future you will gain some idea of any patterns of change in the community.

Each of three groups in the class can be responsible for one aquarium. Members of each group should work out a rota in order to care for, and examine, their aquarium. Sooner or later in the investigation you have just started it is going to become necessary to count the numbers of each type of organism present in the community. This is important because this is the only way to obtain an accurate measure of any changes which may take place. When there is more than one of any type of organism present in a community it is called a population of organisms.

Investigation 3.5 Measuring the size of a population of organisms

Although one way to find the size of a population is to make a complete count of all individuals, this is very often tedious and takes far too long. To save time and effort, samples of the population can be counted and then the total size can be estimated from the samples. This practice is widely used in opinion polls and consumer surveys. When television companies announce that 9 million people watched programme X, they have not counted everybody – it is an estimate based on samples taken in different parts of the country. Why in different parts of the country? Why not concentrate the samples where it is convenient? In the following investigation part a is essential, part b is optional and part c will be performed later with the mini-pond community.

Part a

In this investigation you will measure the size of the population of pupils in your school in order to try out sampling as a method of estimating population size.

Discuss the sampling technique you will use. How many samples will you have and how will you choose them? From the data you collect determine the total school population. The school secretary will know the actual number of pupils in the school. How accurate is your estimate? If you are wide of the mark discuss possible reasons.

How could you extend the ideas you have applied in your school to find out the population of the street or even the town in which you live?

▷Part b

You will need

four wooden laths, 1 m long
large ball of string
two wooden stakes or meat skewers
warm clothes, if necessary

If the weather is suitable, you might like to try estimating the size of the population in one of the communities you examined earlier (see Investigation 3.1). For the sake of convenience, a population in a lawn community is described, but you could easily apply the ideas to other situations.

Select a lawn in which there is a plant which occurs quite frequently but is not the predominant type. Dandelions would be excellent. Work out the total area of the lawn. Discuss how you should take random samples. What should be their size and how many should you count? Estimate the total population of dandelions in the lawn community.

To check the accuracy of your sampling you must find the total number of dandelions. Do this in the following way. The whole class should divide the lawn into a series of lanes, 1 m wide. Mark each lane with string tied to a stake at each end.

The whole class can now concentrate on counting. Divide yourselves into pairs. Each pair takes one lane, and one person counts from each end of the lane towards the middle. When you meet in the middle add your scores together. If you then add up scores for all the lanes you will have a total count done in a relatively short time.

Part c

This should be done for the first time about a month after setting up your mini-pond – although this will depend on the time of year.

You will need

teat pipettes
hand lens
microscope, microslides and coverglass
beaker, 50 cm^3
glass rod

1 Measuring the sizes of the populations in the mini-pond community, such as that shown in figure 3.12, is a much more difficult exercise than measuring the sizes of a population of people or dandelions. For one thing you will be measuring the sizes of several populations at once and there are other difficulties.

Organisms may colonise different places inside the troughs – the open water itself, surfaces such as the sides of the tank or pieces of dead vegetation and so on. It is possible to examine these places for organisms and to count them but this would take more time than you are likely to have. To simplify matters as far as possible concentrate attention on those organisms found in open water.

Stir the water gently in each aquarium. Organisms which are concentrated more in some parts of the aquarium than in others will tend to become distributed evenly. For sampling take several 'dips' with small beakers (50 cm^3 is a convenient size). Try to avoid removing any pieces of dead leaf and so on. Count the number of visible organisms of each type. Do not worry about any proper names throughout this work unless you happen to know them. Invent a name, or give each organism a letter. It is important, however, that you recognise the organism by its given name or letter next time you do this work. Some quick sketches of each organism with its given name will enable you, and other members of your group, to recognise each one next time. Keep a record of the number of beaker 'dips' made because this is the number you must take on all future occasions. If you know the total volume of water in the aquarium you could work out the total population size. This is not essential because we are concerned only in making comparisons of numbers on each occasion you make measurements, in order to see if numbers change.

If the visible organisms are too numerous, or move too fast, you could use the method for estimating numbers described in the table below but substitute the word 'beakers' for 'drops'.

2 Some organisms are too minute to see in a beaker and are often very numerous. It would take a long time to count the actual numbers present in even a small volume of water. Use the following sampling method. Gently stir the water in the aquarium. Take one 'dip' with a small beaker and take it to the laboratory. In the laboratory stir the water in the beaker again. Using a teat pipette remove a few cubic centimetres and place one drop on each of the ten slides and cover with glass slips. Examine each drop in turn with a microscope or hand lens. In the same way as before make a record to enable you to recognise any organism again. Also note in how many of the ten drops each type of organism appears. Record it as rare through to abundant according to the following table:

Figure 3.12
A mini-pond community after many months

Present in 1–2 drops = rare
Present in 3–4 drops = few
Present in 5–6 drops = common
Present in 7–8 drops = very common
Present in 9–10 drops = abundant

Discuss the accuracy of the various methods described in this investigation for measuring the size of populations.

3 Although you are concentrating on measuring the size of populations of organisms found in the open water you may wish to keep a record of those organisms living on the sides of the aquarium and in the dead material and mud accumulated on the bottom. Remove any living material growing on the sides of the aquarium, place it in a drop of water on a glass slide and examine under a microscope. Use a long teat pipette to obtain material from the bottom of the aquarium. Examine a little with a microscope. Ensure that you can recognise the same organisms again. Although this work will tell you nothing about changes in population size it may give you an idea of changes in type of organisms in the community.

4 Keeping a record of observations is an essential part of any scientific investigation and if you are not clear how this is to be done, your efforts may be wasted.

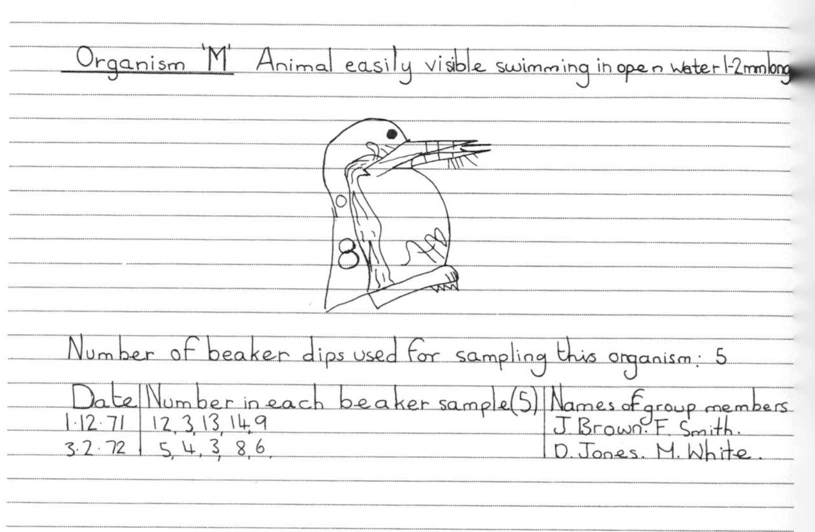

Figure 3.13
A typical page from a log-book kept on a mini-pond community. If your group discovers an organism not previously recorded then you must make the first entry, including the recognition sketch, and decide on the most convenient method of sampling

You need to be able to use your results to do two things easily and fairly quickly:

1 Detect changes in the kinds of organism appearing in the community.
2 See changes in the size of the population of a particular kind of organism.

For these purposes you may find it convenient to record all the results in a single log-book, such as that shown in figure 3.13, kept by the group as a whole. Times and dates are important. Remember that other members of the group depend on you to make accurate records when it is your turn. This means, for example, making a sketch of any new organism and giving it a name or code letter so that it can be recognised by those whose turn it is next.

At intervals information gathered will have to be summarised in the form of tables, graphs or charts. Discuss ways to do this with your group at some convenient moment in the future. Does any pattern of change take place in the community as time goes by? When you are recording and considering results bear in mind the following additional questions. Is there any evidence of:

a Colonisation by both plants and animals?
b Colonisation by animals appearing first, plants later?
c Colonisation by plants appearing first, animals later?
d Certain aquaria containing a greater variety of organisms than others?
e New arrivals replacing earlier colonisers?

Continue to take samples at intervals until you are told to stop. The work will spread over a considerable period of time. Obviously you will need to sample only infrequently in the winter months.

▷ Investigation 3.6 Measuring the size of the human population

From time to time a complete count of the human population of the United Kingdom is made. This is known as a census. The purpose of a census is to find out not merely the total size of the population but also something of the composition of the population and how it has changed since the previous census. The idea is to see if there are any patterns which might be used for predicting future needs so that suitable plans can be made. The diagrams (figure 3.14a and b) show the numbers of people in different age groups in 1891 and 1968.

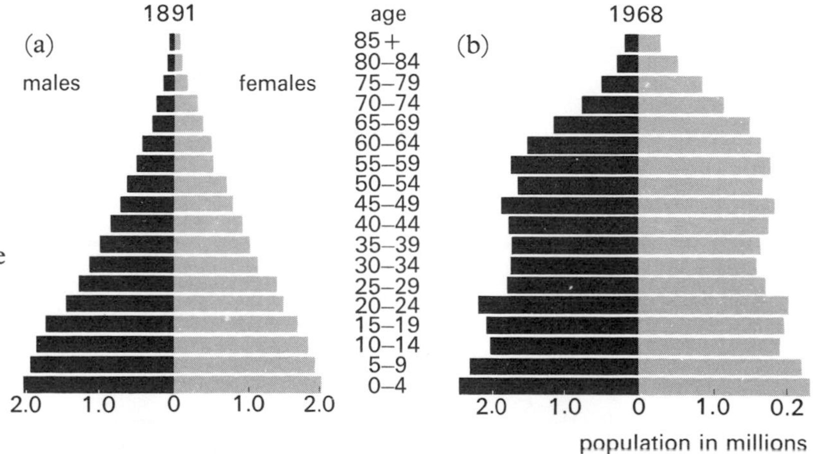

Figure 3.14
The size of the population in the United Kingdom according to age group

1971 CENSUS — ENGLAND

H Form For Private Households

To the Head (or Acting Head) of the Household.

Please complete this form and have it ready for collection on Monday 26th April. If you need help, do not hesitate to ask the enumerator.

The enumerator may ask you any questions necessary to help him to complete or correct the form.

The information you give on the form will be treated as CONFIDENTIAL and used only for compiling statistics. No information about named individuals will be passed by the Census Office to any other Government Department or any other authority or person. If anyone in the census organisation improperly discloses information you provide, he will be liable to prosecution. Similarly you must not disclose information which anyone (for example, a visitor or boarder) gives you to enable you to complete the form.

The legal obligation to fill in the whole form rests on YOU, but each person who has to be included is required to give you the information you need. However, anyone who wishes can ask the enumerator or local Census Officer for a personal form which can be returned direct to the enumerator or local Census Officer and then you need answer only questions B1 and B5 for that person.

PLEASE TAKE NOTE

There are penalties of up to £50 for failing to comply with the requirements described above, or for giving false information.

When you have completed the form, please sign the declaration at the foot of the last page.

MICHAEL REED
Director and
Registrar General

Office of Population Censuses and Surveys,
Titchfield,
Fareham, Hants.

*A household comprises **either** one person living alone **or** a group of persons (who may or may not be related) living at the same address with common housekeeping. Persons staying temporarily with the household **are** included.*

To be completed by enumerator			
C.D. No.	E.D. No.	Form No.	Ref.

If sharing with another household:—

Hall, staircase, passage, etc., shared *only/not only** for entry to accommodation.

*delete whichever is inapplicable.

Number of rooms shared. ....

Name and full postal address:
....
....
....
....

PART A

Answer questions A1—A5 about your household's accommodation and then answer questions B1—B24 overleaf and if appropriate answer questions C1—C7.

Where boxes are provided answer by putting a tick in the box against the answer which applies. For example, if the answer is 'YES': ☑ YES ☐ NO

PLEASE WRITE IN INK OR BALLPOINT PEN

A1
How do you and your household occupy your accommodation?

1 ☐ As an owner occupier (including purchase by mortgage)
2 ☐ By renting it from a Council or New Town
3 ☐ As an unfurnished letting from a private landlord or company or Housing Association
4 ☐ As a furnished letting
5 ☐ In some other way (Please give details, including whether furnished or unfurnished)

....

Note: If the accommodation is occupied by lease originally granted for, or since extended to, more than 21 years, tick 'owner occupier'.

A2
Does your household share with anyone else the use of any room, or hall, passage, landing, or staircase?

☐ YES ☐ NO

A3
How many rooms are there in your household's accommodation? ....

Do not count
Small kitchens less than 6ft. wide,
bathrooms and toilets,
sculleries not used for cooking,
closets, pantries and storerooms,
landings, halls, lobbies or recesses,
offices or shops used solely for business purposes.

Note
A large room divided by a sliding or fixed partition should be counted as two rooms.
A room divided by curtains or portable screens should be counted as one room.

A4
How many **cars and vans** are normally available for use by you or members of your household (other than visitors)? ....

Include any provided by employers if normally available for use by you or members of your household, but exclude vans used solely for the carriage of goods.

If None, write 'NONE'.

A5
Has your household the use of the following amenities on these premises?

a A **cooker or cooking stove** with an oven
1 ☐ YES — for use only by this household
2 ☐ YES — for use also by another household
3 ☐ NO

b A **kitchen sink** permanently connected to a water supply and a waste pipe
1 ☐ YES — for use only by this household
2 ☐ YES — for use also by another household
3 ☐ NO

c A **fixed bath or shower** permanently connected to a water supply and a waste pipe
1 ☐ YES — for use only by this household
2 ☐ YES — for use also by another household
3 ☐ NO

d A **hot water** supply (to a washbasin, or kitchen sink, or bath, or shower) from a heating appliance or boiler which is connected to a piped water supply
1 ☐ YES — for use only by this household
2 ☐ YES — for use also by another household
3 ☐ NO

e A **flush toilet** (W.C.) with entrance **inside** the building
1 ☐ YES — for use only by this household
2 ☐ YES — for use also by another household
3 ☐ NO

f A **flush toilet** (W.C.) with entrance **outside** the building
1 ☐ YES — for use only by this household
2 ☐ YES — for use also by another household
3 ☐ NO

MLC 56-3856 7/70 1

PLEASE TURN OVER TO PART B →

Figure 3.15

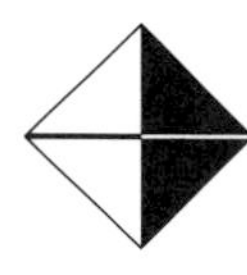

How has the age-group pattern of the population changed in this time? Suggest an explanation. Discuss the social consequences of the changes in the pattern between the two dates. Is it possible to predict social consequences in the future? Look at the numbers in the 20–24 age group in figure 3.14b. You can see why this group is known as the 'post-war bulge'. ▶Predict the size of the next 0–4 age group.◀ How certain are you about this prediction?

On the basis of the probable prediction, what plans need to be made for the future (for example, between 1973 and 1978)? What could happen if the prediction were wrong? Discuss the differences between making predictions from patterns such as this and that of 'acid and carbonate produce carbon dioxide'.

The 1971 census showed that the population in England and Wales had risen from nearly 44 million in 1951 to 48.6 million in 1971. The growth of the human population in this country and the world is seen by many as a very serious problem.

THE POPULATION CRISIS

pill

Government told to control 'intolerable' rise in population

World moving towards maximum population figure

Doctors want commission on overpopulation

An average completed family size of 2.1 children would give us a stable population

longer life promised

Figure 3.16

The headlines collected from newspapers and magazines and shown in figure 3.16 indicate the concern felt by many people. Discuss the reasons why there is concern about the increasing size of the human population. You may or may not agree with the concern as expressed by many people but to help you form an opinion it would be of great value to have some knowledge of the ways in which populations grow in size. It may then be possible to make predictions about the size of the population in the future. It may also be of value in helping you to understand any patterns of change observed in the mini-pond community.

Investigation 3.7 The growth of populations

You will need

For part a
large number of small beads (matchsticks or any other suitable objects would do)
graph paper
For part b
cultures of *Drosophila*
ether and emergency etheriser
sheet of filter paper
mounted needle

Part a

The purpose of this investigation is to observe the way in which a population might grow by using beads to represent the organisms. Use the beads to represent a real population – say mice on an island. Begin with a population of six mice – three males and three females. The assumptions to be made in order to simplify matters are:

a Each pair of mice produces ten offspring each year, five females and five males.
b Each year all the parent mice always die.
c Each year all the offspring survive to breed.
d During the time of study no mice leave or reach the island.

How do these assumptions probably differ from the real situation? Using the beads or matchsticks to represent organisms and bearing the assumptions in mind, work out the numbers of organisms in the population for each of the next six years.

It is, of course, quite possible that you will not need beads or matchsticks as a help; you may prefer simply to calculate these figures. Construct a graph with years on the horizontal axis and numbers of organisms on the vertical axis.

Just as we need tools like a telescope to help us extend our powers of observation, so we need 'mental tools' to help us extend our thinking. One such mental tool is called a model.

When scientists use the term model they do not mean a smaller version of something much bigger. Such smaller versions are often as accurate and detailed as the real thing. Scientists use the term model to mean something which helps them understand, or explain in a simplified way, real and complicated situations. Sometimes models are physical; in other words they are actual objects which represent and perhaps work in the same way as the real thing. You are using one such model in this investigation. Because models are simplifications, they differ in some respects from the real situations.

The simplifications we make are called assumptions. To simplify, we assume certain things that may be quite correct, or may be only approximately true; or we may assume that they are not necessary to the immediate purpose for which we use the model. It is important to keep such assumptions in mind when we use a model, for if it is later found to be incorrect these assumptions can easily be checked. By using assumptions we are merely making the work easier – not accepting them as facts.

If you did not see the beads but simply calculated the figures, you have taken your 'model-making' a step further, so that you are now using a mathematical model of population growth.

The next investigation you perform will be to see how closely the growth of a real population matches the model. To do this you will study some experimental populations and also examine the data in parts b–e.

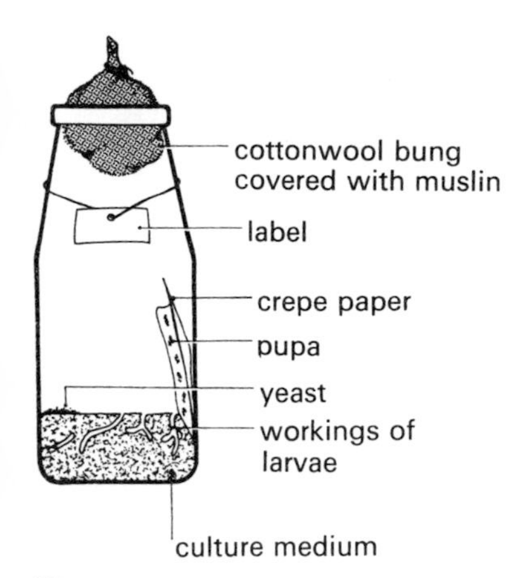

Figure 3.17
A *Drosophila* culture bottle

Part b

The fruit fly *Drosophila* is a convenient organism to use because it is easily kept in the laboratory, its population grows fairly rapidly and it is easy to count. You will either commence this investigation now, to be completed in several weeks' time, or you will be given a series of populations of *Drosophila* which were started at intervals some time in advance. The population dated most recently will therefore be the youngest. If the investigation is commenced now, part c and Investigation 3.8 will be left until it has been completed. When you have made a series of counts of the population size over a time period, draw a graph with numbers on the vertical axis and time on the horizontal axis.

How does the growth in size of the population of *Drosophila* compare with the model population? Discuss ways in which the real population of *Drosophila* probably differs from the model population.

Part c

Further data on the growth in size of populations.

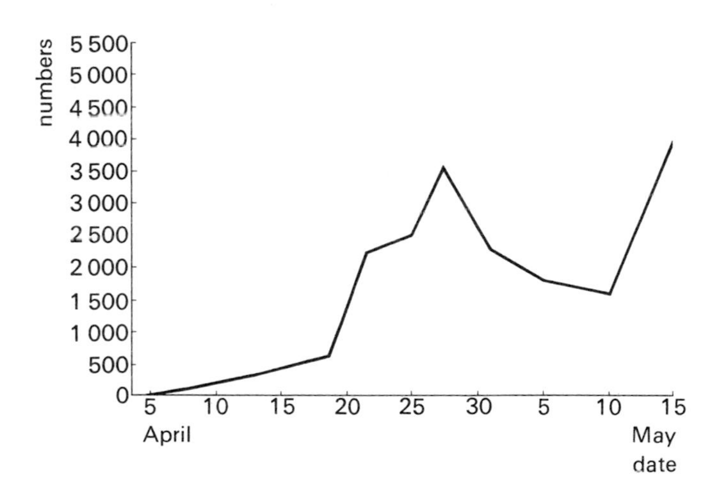

Figure 3.18
Changes in the size of a population of mites in an artificial laboratory situation

Figure 3.18 shows the growth in size of a population of mites in an artificial, laboratory situation. The *Drosophila* in part b were also grown under artificial conditions (see figure 3.17). Discuss ways in which this may affect the results obtained.

The following tables give figures showing the growth of a population in a colony of bees, and the growth of the human population in England and Wales.

Bees

days	0	7	14	21	28	35	42	49	56	63	70	77	84	91	98	105	112	119
population (in thousands)	1	1.5	2.5	4	8	16	22	32	40.5	50.3	55	62.5	72	72.5	71	82	78	81

Man

date	period	population	
100000 BC	Old Stone Age	a few hundred	
10000 BC	Middle Stone Age	a few thousand	
1400 BC	Bronze Age	100000	see note 1
400 BC	early Iron Age	150000	
250 BC	late Iron Age	240000	
AD 200	Roman occupation	700000	
AD 400–900	Dark Ages	500000	
AD 1086	Domesday book	1500000	see note 2
AD 1340	before the Black Death	3900000	
AD 1377	after the Black Death	2360000	
AD 1430		2200000	see note 3
AD 1545		3470000	
AD 1695		5200000	
AD 1750		6200000	
AD 1801		8900000	census figures
AD 1861		20100000	
AD 1901		32200000	
AD 1941		42200000	
AD 1961		46100000	
AD 1971		48600000	

1 These figures are guesses.
2 The Domesday Book provides an excellent source for determining the population of parts of England but it does not cover the whole of England and Wales. Although the Black Death took an appalling toll directly on the population, there were other factors which contributed to the reduction; for example, successive crop failures.
3 During this period parish registers provide a useful source for calculation of the population size.

Is there a pattern in the growth in size of populations?

▷Part d

In the previous parts of this investigation you were asked to draw graphs with numbers on the vertical axis and time on the horizontal axis. However, it is quite probable that the numbers became so large that it was impossible to fit them on a single piece of graph paper. One way of overcoming this problem is to use special logarithmic graph paper on which the lines are not equally spaced. Using normal graph paper, the gap between 200 and 400 is twice the size of the gap between 100 and 200. This is obviously because every 100 has an equal value wherever it is put on the axis. Using logarithmic paper the effect is different because the gaps between 100 and 200, 200 and 400 and 1 000 and 2 000 are all of the same size. This method of drawing the graph allows much greater numbers to be included. Use the logarithmic graph paper to draw graphs of the growth of some populations. What effect does this have on the shape of the line drawn to illustrate population size? Which type of graph gives the impression of less dramatic growth in size of a population?

▶Part e

Predict the population of England and Wales in the year AD 2000.

In part a it was necessary to make some assumptions before using the model population. What assumptions is it necessary to use in making this prediction? How certain are you that your prediction is correct?◀

> The population of the United Kingdom will, according to the latest official projection, grow by some 11 million, which is a fifth, between now and the end of the century. The projection is continually being revised, and the end of the century figure has in fact been reduced by eight and a half million within the space of five years. The latest projection amounts to an increase of half per cent a year over the next 30 years, modest by the standards of most of the rest of the world.

The Times (19 May 1971)

Suggest reasons for the continual lowering of the estimated size of the population in the year 2000. How confident do you feel about your own predictions?

However, in a report on the population growth of the United Kingdom, a special committee set up to recommend action to the Government stated:

> Despite margins of error the cardinal [very important] fact remains that all recent projections agree on a substantial and continuing increase in the population.

Despite the difficulties of accurately predicting the future size of the population it would seem that some increase is sure to happen. Compare the patterns of growth in population size shown by man, your experiments and the other data. Predict what could possibly happen to the growth of the human population eventually. Can you be sure about your prediction? If not, why not?

Making predictions about populations and their future trends is a very difficult and complicated subject. It is impossible to be precise because not enough information is available on which to base predictions. Nevertheless, you saw at the beginning of this investigation that the present size of the human population and its possible future growth is, for many, a cause for grave concern. For various reasons others are less concerned. Because it is such a complicated subject, there will be opportunities to discuss the problem of population growth in other parts of the course, when you have discovered more patterns on which to make judgments!

Investigation 3.8 Growing populations together

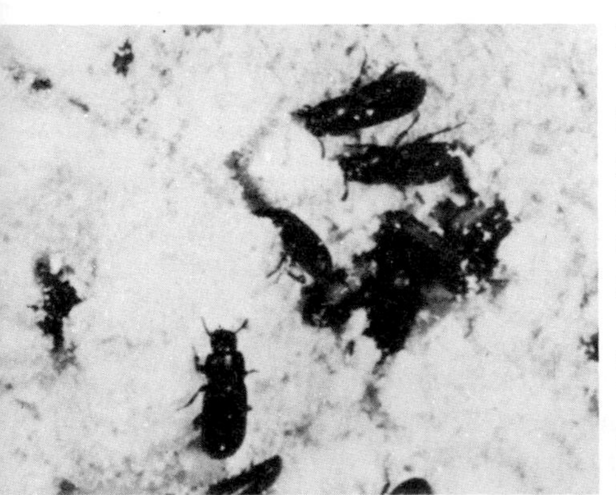

Figure 3.19
The flour beetle, *Tribolium sp.*
The types of flour beetle used in Investigation 3.8 are very similar

Initially this is a demonstration investigation.

One of the reasons for studying populations was to enable you to measure, and possibly explain, patterns of change in a mini-pond community. It was easiest to measure the growth in size of a single population. However, it would be more like the actual situation if you could see what happens when two populations are grown together.

You already have enough to do, so this investigation will be set up and looked after for you. Two similar types of beetle called *Tribolium confusum* and *Tribolium castaneum* will be used. Both are serious pests of stored grain and flour in hot countries. The results will be examined later.

4 Looking at organisms

Figure 4.1 How big?

If you were asked to describe an elephant one thing you would almost certainly say is that elephants are big. But to a whale an elephant is small; to a mouse a man is big. You describe things as you see them: you judge them according to your own size. In order to avoid using words like big or small, enormous or microscopic, which give only a general and rather vague description of size, man has developed instruments which enable him to measure size accurately and consistently. You can read more about this in the book *Length and its measurement.* In this section you will begin by looking at, and measuring, the sizes of different organisms. Are there any limits to the size of organisms? Are there any obvious, but interesting, patterns?

Investigation 4.1 Measuring the size of organisms

You will need
microscope
transparent ruler with a scale marked in millimetres
prepared microslide with any small specimens
culture of *Paramecium*

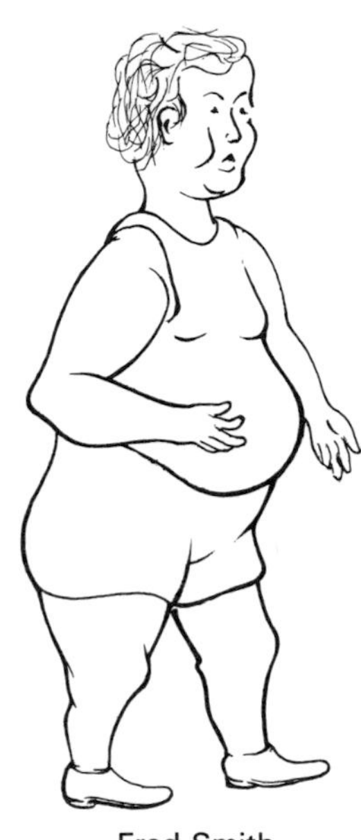
Fred Smith

Figure 4.2a

Joe Brown

Figure 4.2b

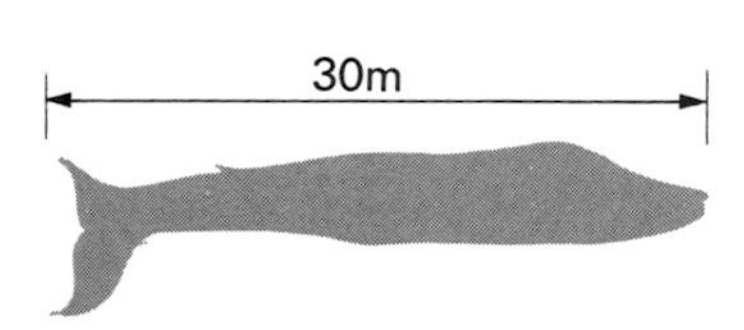

blue whale – the largest animal that ever lived

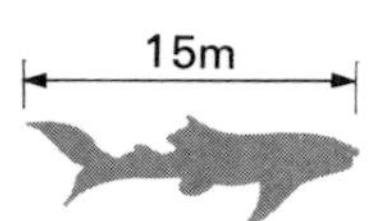

whale shark – the largest fish

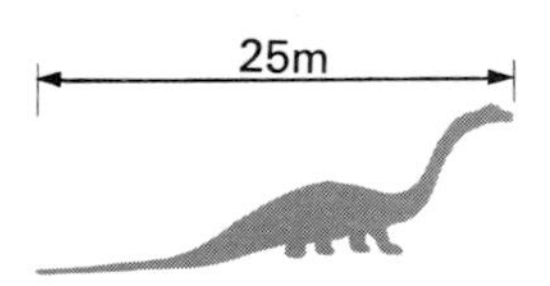

Brontosaurus – extinct, giant swamp dwelling dinosaur

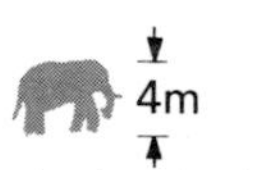

elephant – largest land animal

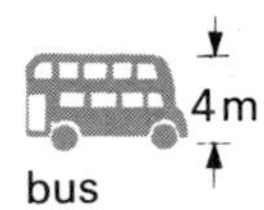

bus

6m

giraffe

Baluchitherium – largest land animal that ever lived, now extinct

Figure 4.3
Are the organisms shown drawn to the same scale? If so what size would you draw yourself?

Look at figure 4.2. Fred Smith and Joe Brown are both 1.75 m tall, but does this tell us much about their overall size? It is clear that other measurements must be made if an accurate idea of their true size is to be obtained. What other reliable measurements of their size could you make?

Consider and discuss ways of measuring the size of some of the organisms normally kept in your laboratory.

Measuring the size of very big, or very small, organisms can pose special problems. How, for instance, can the size of organisms too small to be seen with the naked eye be measured?

Measuring the size of the very small

A microscope (see figure 4.4a–b) is an essential instrument for investigating small things (and for seeing small details of large things!). If you have not used a microscope before you will be shown how to operate one.

In order to measure the size of very small objects you must have some idea of the size of the field of view you have when you look down the microscope. This can be done in the following way or you may be told to use an alternative method.

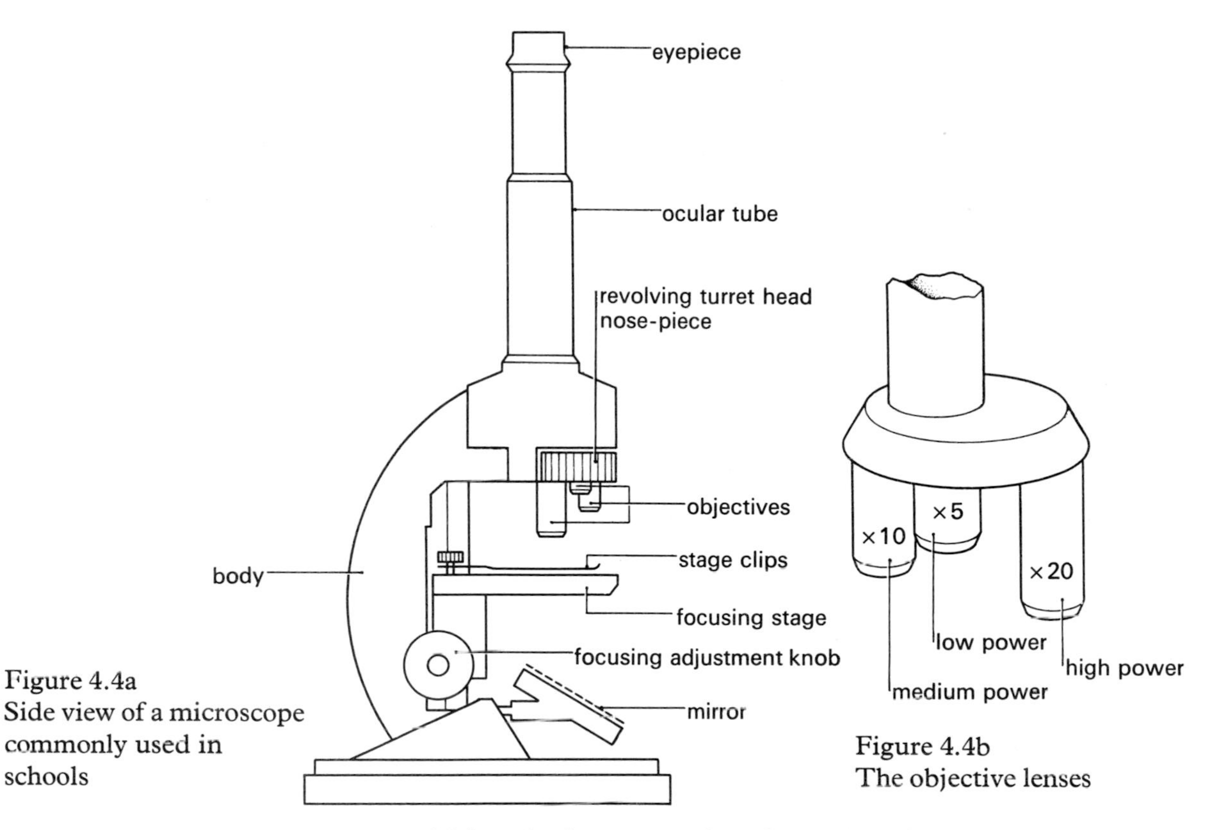

Figure 4.4a
Side view of a microscope commonly used in schools

Figure 4.4b
The objective lenses

Using the low power lens focus the microscope on the 1 mm marks of a transparent plastic ruler as shown in figure 4.5. What is the diameter of the field of view of your microscope? What happens to the field of view if you replace the low power lens with one of higher power? What is the diameter of the field of view now?

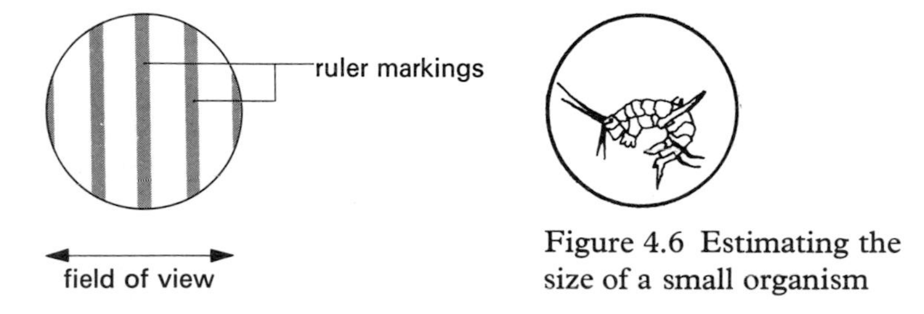

Figure 4.5
A transparent ruler in the low-power field of view of a microscope. The markings are 1 mm apart. What is the diameter of the field of view shown?

Figure 4.6 Estimating the size of a small organism

Figure 4.6 shows a small insect seen through the microscope. The field of view is the same size as that in figure 4.5. What is the size of the insect?

Of course most of the organisms, or parts of organisms, that you wish to measure will be awkward portions of a millimetre. If you are familiar with indices you will know that one way of writing a millimetre is:

10^{-3} m (one thousandth of a metre: $\frac{1}{1000}$ metre).

One tenth of a millimetre is:

10^{-4} m (one ten thousandth of a metre: $\frac{1}{10\,000}$ metre).

One thousandth of a millimetre is:

10^{-6} m (one millionth of a metre: $\frac{1}{1\,000\,000}$ metre).

For your present purposes units of 10^{-3} m are probably satisfactory. Rather than writing the size of an organism as 1.2 mm it would be better as 1.2×10^{-3} m.

Measure the length of the small organisms which you will be given.

One simple pattern which this investigation has emphasised is that there is an enormous range in the size of organisms. But you were probably quite well aware of this, and anyway the main purpose was to look at different ways of measuring the sizes of organisms. Are there more interesting, and perhaps more useful, patterns about the size of organisms?

▷ Investigation 4.2 Investigating patterns in the size of organisms

In measuring the length or height of organisms you were making linear measurements. What other linear measurements of organisms could you make? Sometimes, however, linear size measurements –

as in the case of Fred Smith and Joe Brown – do not really give a very good idea of size. You probably discussed this and may have considered the possibility of using mass as a measurement of size. It would be easy to measure the mass of some of the organisms in your laboratory.

The figures in the following table refer to the approximate masses of a variety of animals. Their habitat (where they live) is also given.

organism	approximate mass/kg	habitat
Indian rhinoceros	1 700	land
dog	13	land
Brontosaurus⋆	40 000	swamps
elephant	4–5 000	land
Baluchitherium⋆†	6–7 000	land
horse	700	land
man	55	land
brown bear	550	land
blue whale	120 000	sea

⋆Extinct, so the figures given for mass are guesses.
†The largest true land dwelling animal ever known.

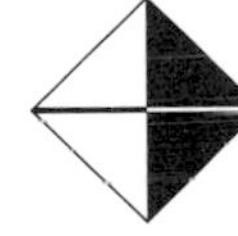

What pattern does the table show? ▶What is the explanation? The photographs in figure 4.7 should give you a clue.◀

Examine the skeletons in figure 4.8. What pattern do they show? ▶Suggest an explanation for the pattern.◀

Figures 4–7
Jellyfish float and swim in open water

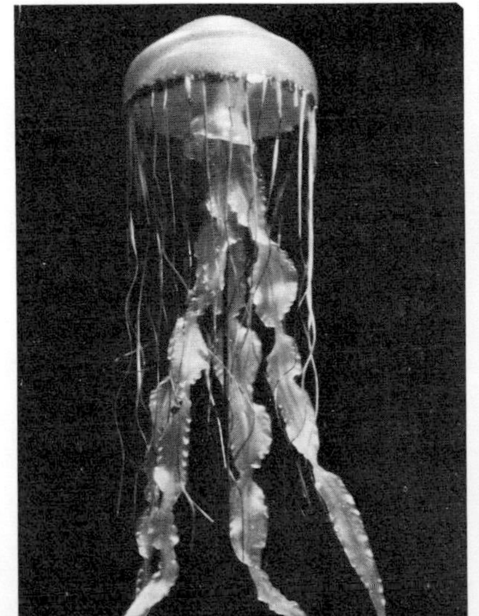

Contrast the appearance of these stranded jellyfish with that of the one in the water.

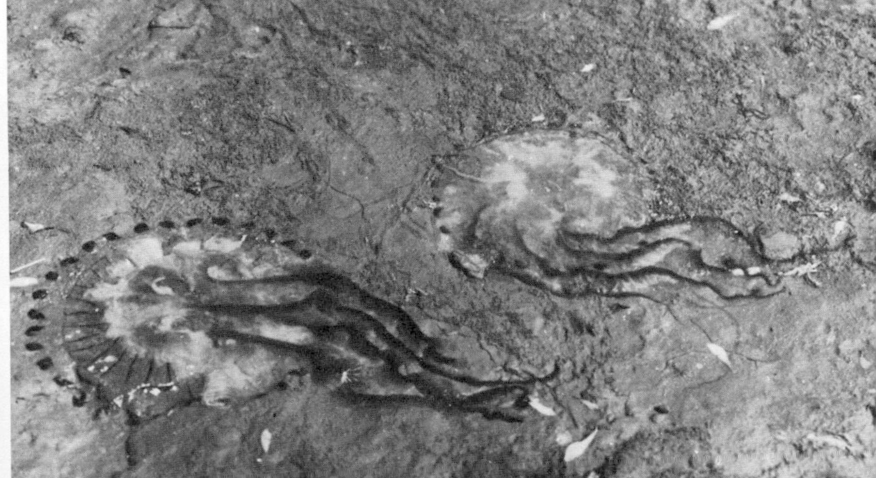

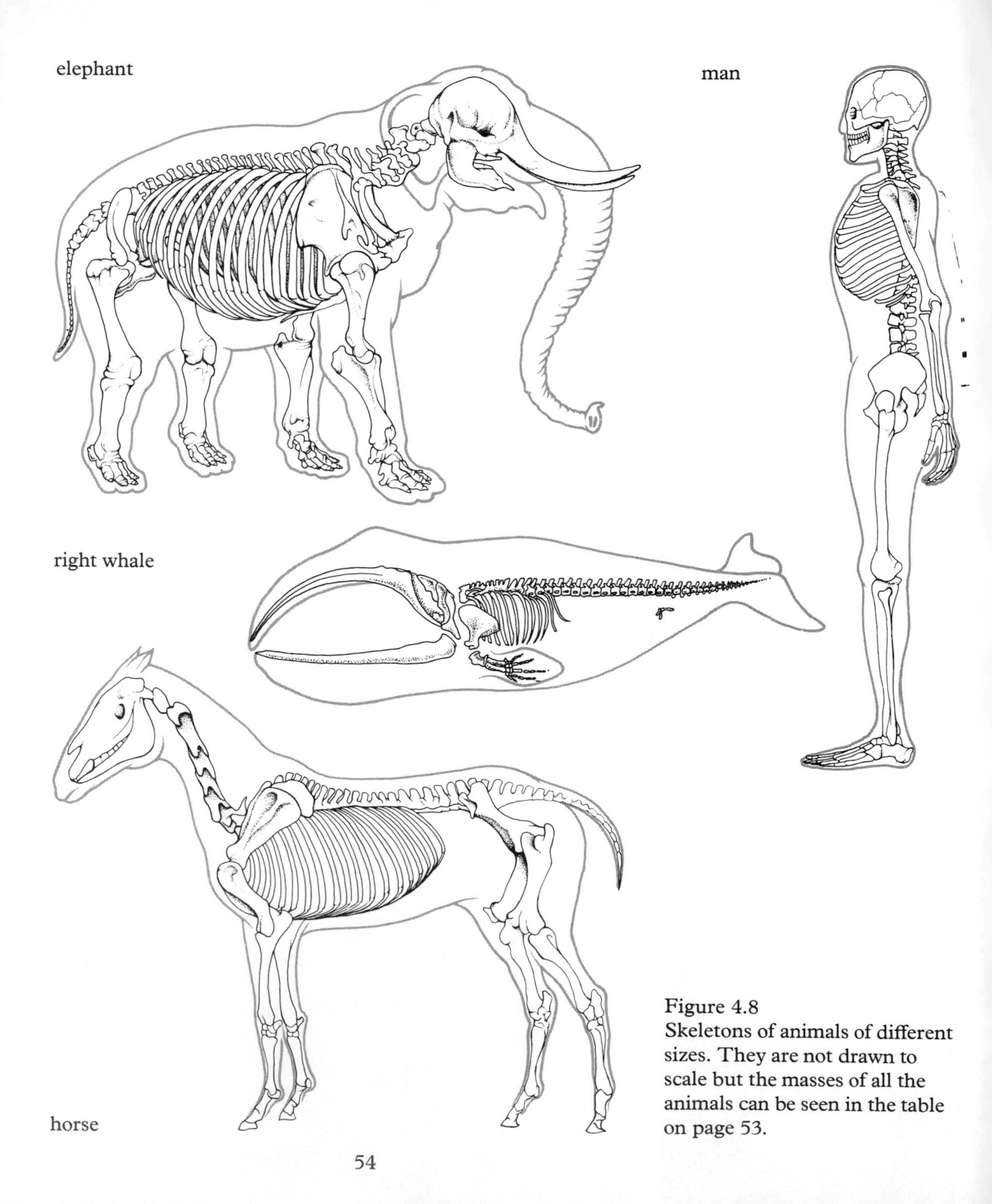

Figure 4.8
Skeletons of animals of different sizes. They are not drawn to scale but the masses of all the animals can be seen in the table on page 53.

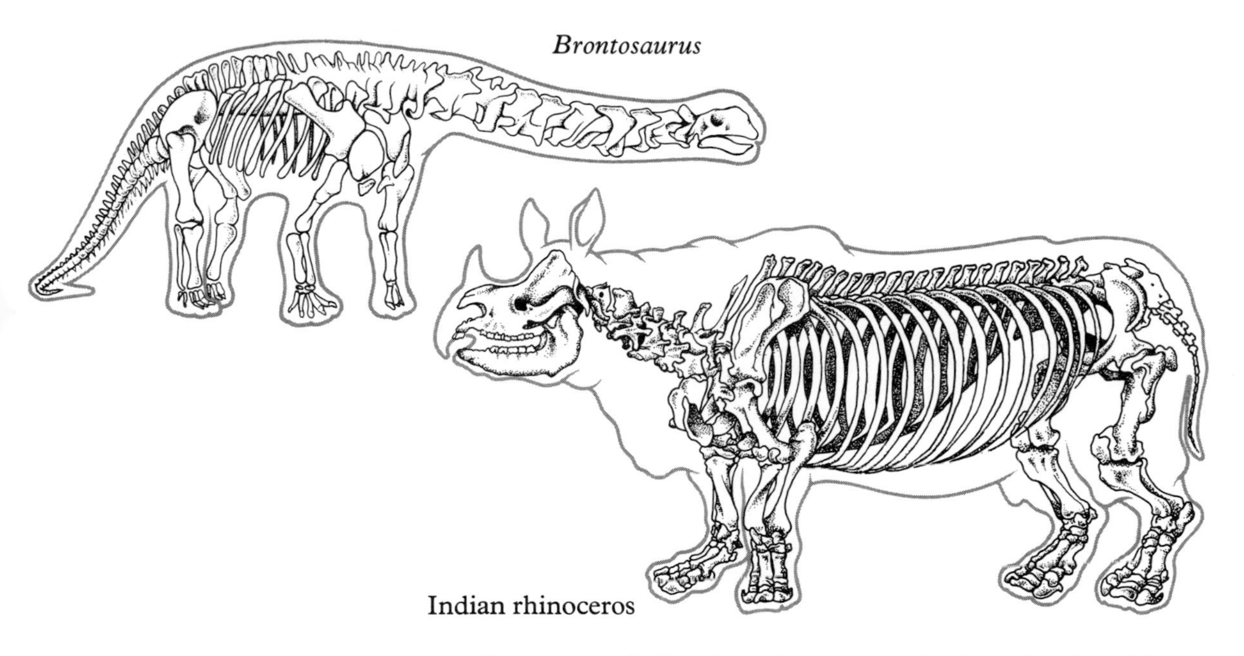

▶Do you think that there is an upper limit to the size of land-living animals? If so, why?◀

Fred Smith and Joe Brown showed you that when thinking about the size of organisms there are several types of measurement which should be taken into account. It would be interesting to compare the various measurements of size as organisms grow.

Investigation 4.3 Comparing measurements of linear dimension, mass and area as organisms grow

You will need

thirty-two small wooden cubes each with all sides 1 cm long

In this investigation you will use a wooden cube as a model organism. It is possible that you will not need an actual model. You may be able to use a mental picture instead, but you will still need to measure the mass of a number of cubes in order to find the average mass of a single cube. What are the volume, the mass and the total surface area of a single cube? Make the model organism grow by doubling all linear dimensions. What are the volume, the mass (since the model organism is made of the same 'stuff' all through you can work this out) and surface area of the larger organism? Now make the model grow again by adding another 1 cm to all the sides and repeat the calculations. Carry on making the model grow. You may find it easier to place your calculations in a table:

linear dimensions of each side/cm	volume/cm^3	mass/g	total surface area/cm^2
1	1		6
2			
3			
4			

What is the pattern connecting the volume and mass of the model organism as it grows? Remember it is only a model organism. Why would you have to be cautious about accepting the same pattern for real organisms?

What is the pattern connecting volume (and mass – at least for the model) and total surface area as the model organism grows? You may find this easier to appreciate if you find the number of cm^2 of surface area for each cm^3 of volume (or g of mass) by dividing the surface area by the volume.

▷ Investigation 4.4 The influence of shape

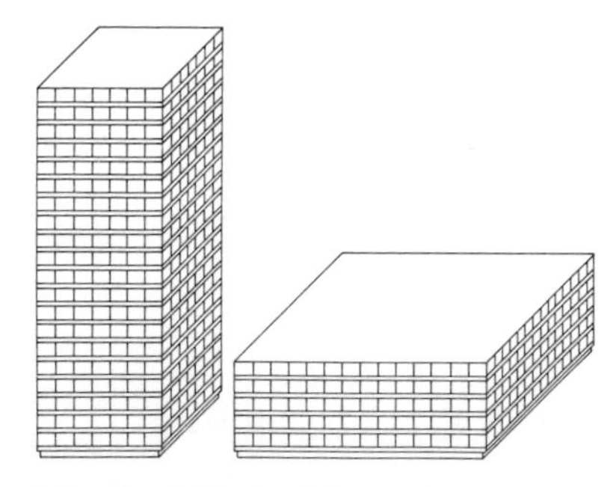

Figure 4.9a and b

When architects are asked to design a new building usually they are given some limits within which they must work. For instance, local planners may set a maximum height for buildings in their area. Imagine you are an architect given the task of designing a new building. It must be less than the maximum height allowed so that no more than 20 floors can be included. You are given a limit to the amount of money you can spend and one way of economising is to keep very small the amount of expensive external finishing materials used.

▶ If this was more important than anything else would you design the building like figure 4.9a or 4.9b assuming the floor area is the same in both cases? ◀

What other considerations are important in designing buildings?

▷ Investigation 4.5 Can the pattern be extended to real organisms?

You will need

sheets of newspaper

variety of animals, e.g. gerbils, rabbits, hamster, you, teacher
apparatus to measure organisms of large and small masses

►Predict the connection between surface area and mass of the organisms you have available. Even if you ignore shape, what major assumption will you probably need to make? Use the materials supplied to test your prediction.◄ Is it precisely correct, roughly correct or incorrect? Discuss your results in view of the fact that the pattern used for prediction was obtained from a model organism and has been applied to real organisms.

One of the very first illustrations (figure 1.1a) in this book was of a number of different animals. You were asked if you could see any patterns about them. One pattern which you could not see from the pictures, but which you might know, is that they are all warm-blooded. What other group of animals is warm-blooded? One of the problems for warm-blooded organisms is that they must keep warm, otherwise they will soon die of exposure. This happens often to people who do not make sure of keeping warm in extreme conditions. What features of organisms help to keep them warm? Apart from the obvious features could size patterns have anything to do with it?

Figure 4.10a Keeping warm

Figure 4.10b Keeping cool

Investigation 4.6 An investigation into keeping warm and cooling off

You will need

you must decide for yourself what apparatus you will need for this investigation

Figure 4.11
Map showing the distribution of penguins. An average temperature for each habitat and average sizes of each type of penguin are given in the accompanying table.

Study the information given in the following table and figures 4.11 and 4.12 about penguins. These birds are interesting because they are all more or less the same shape (why?) but their size varies considerably. You therefore need only be concerned about size (not shape); in this way they are similar to the cube model. You can assume that all penguins are made more or less of the same materials no matter what their size, again like the model.

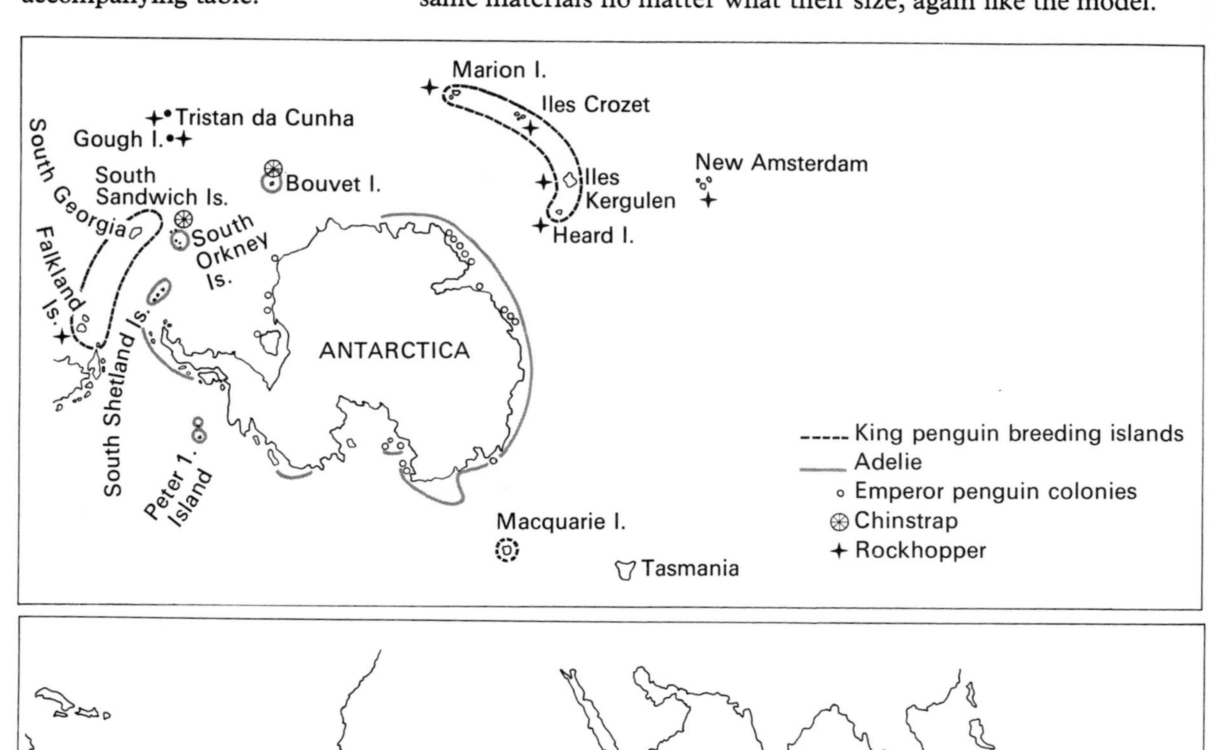

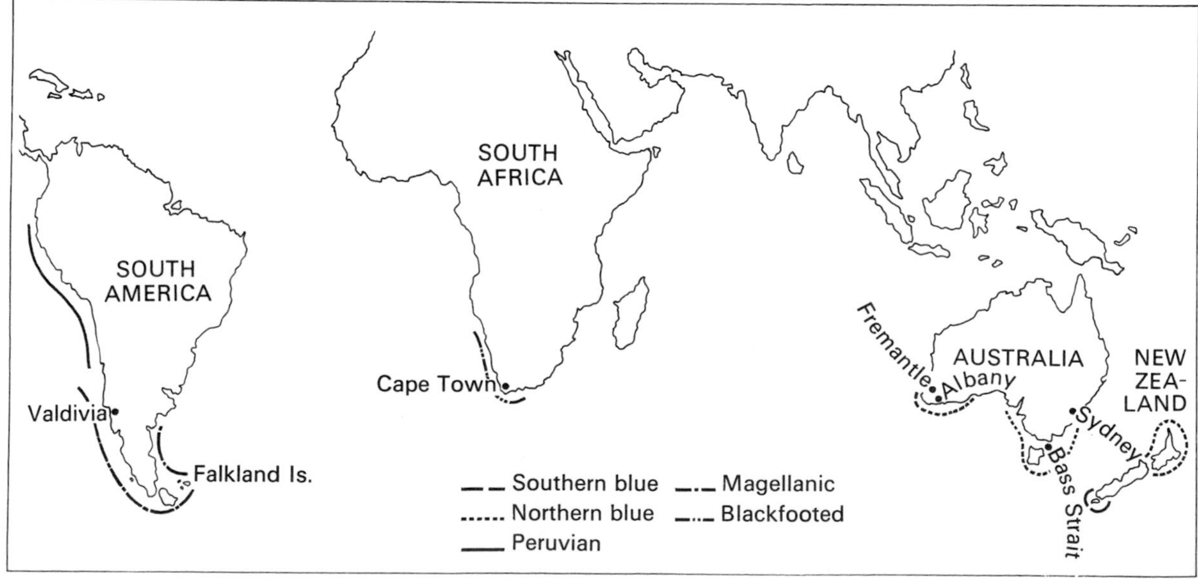

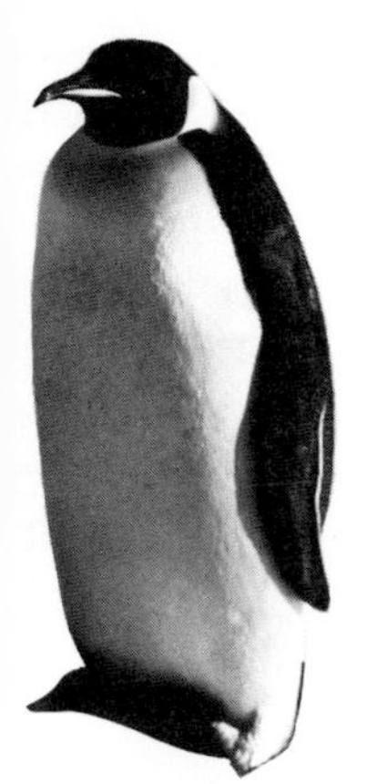

Emperor penguin

Figure 4.12

penguin	mass/kg	approximate temperature range of habitat(s)/ °C
Emperor	29.15	−10 to 0
King	16.0	0 to +7
Adelie	5.0	−10 to +5
Magellanic	5.0	0 to +12
Chinstrap	4.5	−5 to +5
Peruvian	4.3	+6 to +16
Rockhopper	2.8	+5 to +10
Blackfooted	2.8	+10 to +17
Northern blue	1.8	+12 to +19
Southern blue	1.8	+8 to +15

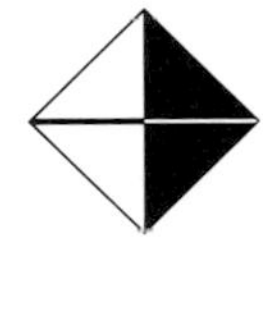

What pattern is clearly shown by the information? ▶Suggest an explanation for this pattern. Devise a way to test your explanation using ordinary laboratory apparatus as models.◀

You have seen that different types of organism (such as penguins) vary in size, but it must also be obvious that individuals of one type of organism also vary, not only in size, but in many other ways. You only need look at the members of your class to appreciate this simple pattern! However, there are two interesting questions which arise out of this:

a Has the variation any pattern?

b Why is it that individuals of a type of organism vary?

You will attempt to find an answer to the first question and at least a partial answer to the second.

Peruvian penguin

Investigation 4.7 Searching for patterns of variation

You will need

pupils in the year group at school
apparatus to measure their heights
graph paper

It is suggested below that you investigate three different features for patterns of variation. There is no need for you all to investigate each one. The class can be divided into two groups. Pairs from each group can then gather data to pool with other members of the group later.

1 Height

Measure the height without shoes (to the nearest centimetre), of

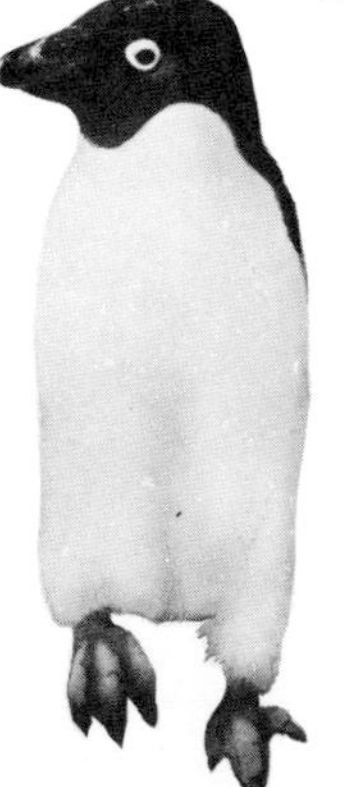

Adelie penguin

each person. If you are in a mixed school record the heights of boys and girls separately. Make as many measurements as you can in the time available.

Construct large histograms of:

a all the boys together
b all the girls together
c everybody together.

What is the pattern of variation in height shown by the histograms? Why did you measure persons only in the same year group at school?

2 Tongue rolling and
3 Ear lobes

You can investigate (2) and (3) at the same time.

The photographs in figure 4.13a and b show the two variations under investigation. Search for a pattern concerning both these variations amongst your fellow pupils. Gather as much data as you can to back up any suspected pattern.

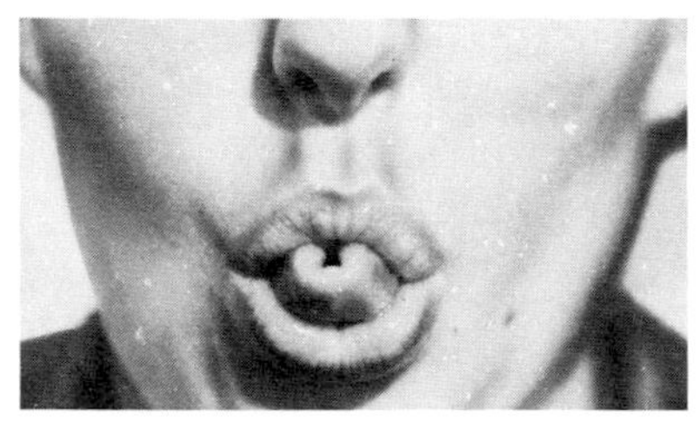

Figure 4.13a
Tongue-rolling ability

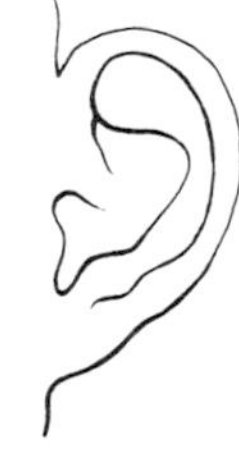

(i) attached

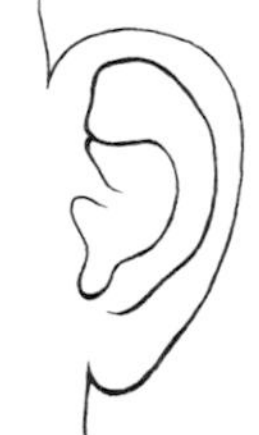

(ii) unattached

Figure 4.13b Ear lobes

Compare your results with those obtained by the other group. Discuss the patterns of variation obtained.

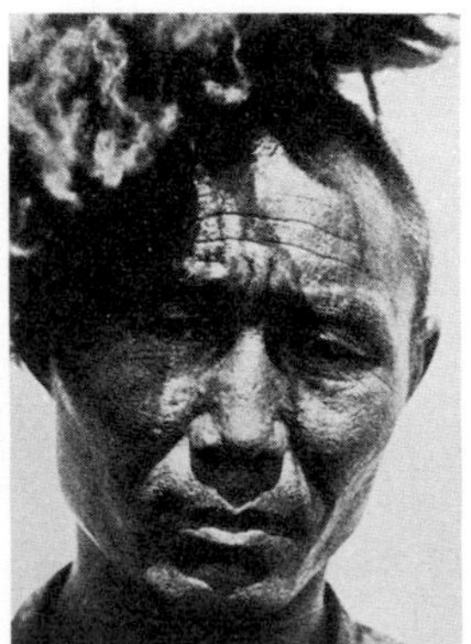

Figure 4.14
Some variations of man

In this Investigation have you really been concerned with the organism building block or with a different building block?

The races of man (see figure 4.14 for some examples) show a great many variations. Discuss some of them and their significance.

▷ Investigation 4.8 A further look at variation patterns or 'How to be honestly dishonest!'

You will need

graph paper

Answer Test 1, giving yourself one mark for each 'yes' and no marks for any other answer.

Test 1

1 I would rather read a story about horses than a space story.
2 I would prefer lessons on using make-up rather than experimenting with electronics.
3 I prefer to play tennis rather than play rugby.
4 I would prefer to go to a mixed party rather than to a party for people of just my own sex.
5 I would quite like my bedroom to be painted pink.
6 I like to take babies for a walk.
7 I think puppies are nicer than big dogs.
8 I would like to bathe in a bubble bath.
9 I would rather wear a dress or skirt to school than trousers.
10 I would rather be a fashion model than a civil engineer.

If there are both boys and girls in your class, treat results for boys separately from those for girls. Draw a histogram with the number of persons on the vertical axis and the number of 'yes' answers on the horizontal axis. Two histograms will be drawn for a mixed school.

What is the pattern of variation in 'yes' answers? Comment on the questions in the text. Were they fair? Now answer Test 2 and draw another graph.

Test 2

1 I regularly read library books.
2 I like experimenting with chemicals.
3 I enjoy PE and games.
4 I prefer the company of one or two friends rather than attending a party.
5 I like dull colours.
6 I like to go for walks.
7 I would like a dog at home.
8 I really enjoy taking a bath.
9 I usually wear bright-coloured clothes.
10 I would quite like to be a teacher.

What is the pattern of variation in 'yes' answers?

Compare the questions in Test 2 with those in Test 1. You can see how it is possible to phrase a question in such a way that the patterns of variation in 'yes' answers can be contrived. Look for examples of this in advertising, in newspapers or in TV interviews.

You can now consider the second question which was posed earlier: Why is it that individuals of an organism vary?

Investigation 4.9 Why are you like you are?

'Oh, isn't he like his father!' You must have heard someone say this many times about a new baby. Perhaps you are like your mother or father. It is an obvious pattern that children resemble their parents. Can the pattern be extended to other organisms?

The variations which add up to you being you were obviously inherited from your mother and father; but is this all? It is necessary to look more closely at the reasons for your being like you are.

Study the information given about twins. There are two types, the so-called identical twins, and fraternal or non-identical twins.

You probably know that you began your life when two gametes, the egg and the sperm, united to form a zygote from which you grew. You can find out more in the book *Patterns of reproduction, development and growth.* Study figure 4.15.

Fraternal twins have a similar start in life to ordinary brothers and sisters. The only difference is that they grow at the same time in their mother's womb. Identical twins grow from the same zygote; they share the same sperm from the father and egg from the mother. Fraternal twins, like any other children from the same parents, grow from different eggs and sperm.

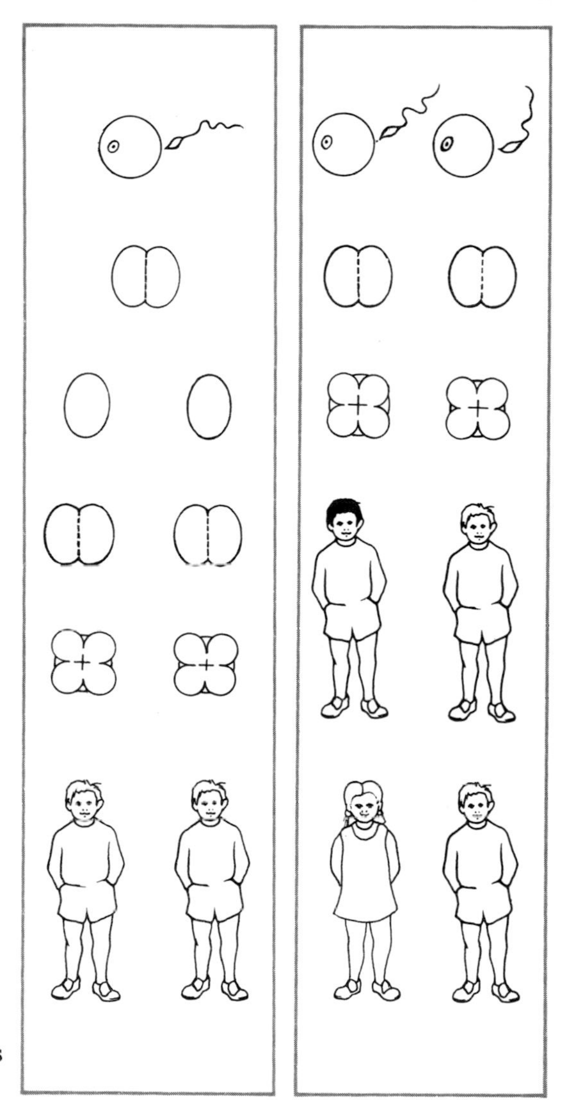

Figure 4.15
How identical twins and non-identical (or fraternal) twins come about

Data: Set 1 – Identical girl twins (A and B) compared with a sister (C)

characteristics which may vary	A	B	C	Mother	Father
hair colour	fair	fair	fair	brown	fair
eye colour	blue	blue	hazel	blue	hazel
ability to roll tongue	no	no	no	yes	no
ear lobe attached	yes	yes	no	yes	no
freckles	yes	yes	yes	yes	no

Figure 4.16 These identical twins were separated at birth and brought up apart

Are these data what you would expect? If you have identical twins in your class you could get similar information about them if they agree.

Data: Set 2 – Differences in physique between members of a pair of twins and ordinary brothers and sisters. These differences are averages taken from many sets of twins and ordinary brothers and sisters

characteristics which may vary	identical twins brought up together	fraternal twins brought up together	ordinary brothers and sisters brought up together	identical twins brought up apart
height/cm	1.7	4.4	4.5	1.8
mass/kg	2	4.5	4.6	4.5
head length/cm	0.29	0.62	–	0.22

From an examination of these data what do you suspect is responsible for human variation apart from inheritance? From everyday experience discuss further evidence which would support this view. Can you extend the pattern which suggests that variation is the result of what you inherit from your parents? ▶Why are you like you are?◀

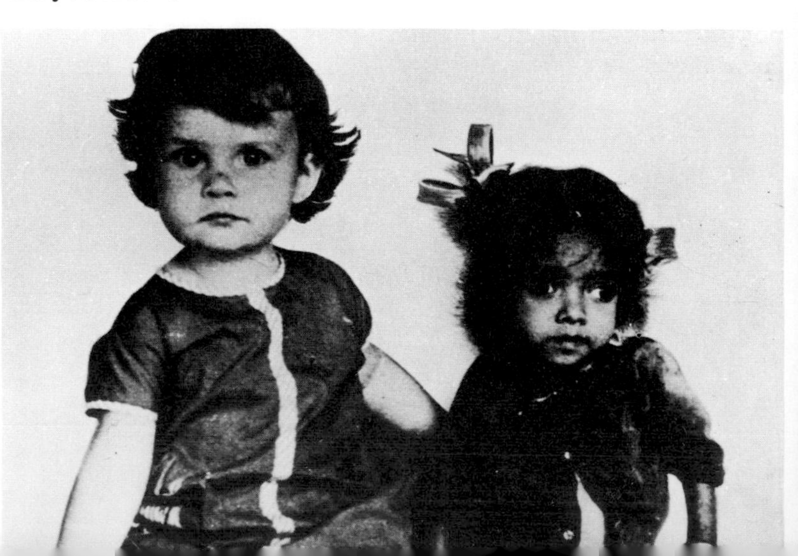

Figure 4.17
The Indian girl is three years old. She is seriously retarded, mentally and physically, as a result of a lack of good food throughout her life. In contrast the other girl is only eighteen months old

Although the pattern may be extended to include environmental influences, can it be extended for other organisms? Once again you should be able to discuss the possibilities arising from everyday experience.

Figure 4.18 A tree growing in an exposed position compared with a similar type of tree growing in a more sheltered place.

In the example shown in figure 4.18 – and probably in others you may have seen or discussed – there is a reason for objecting to it as evidence for the extension of the pattern you have just discovered. What is this reason? Outline the plan of an investigation you might perform which would overcome the objection.

▷ Investigation 4.10 Inheritance, environment and variation

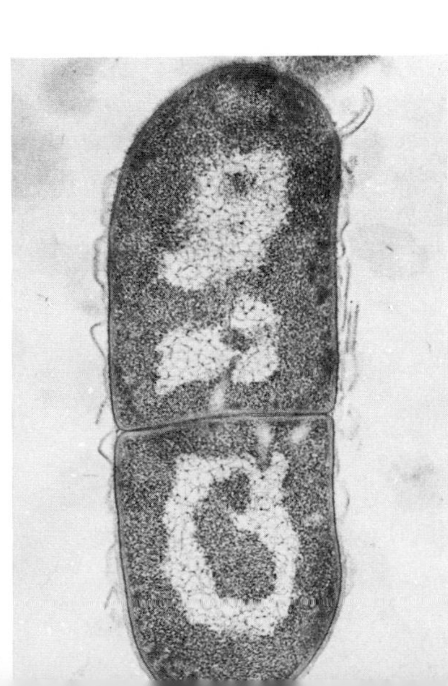

Figure 4.19
Photomicrograph of bacteria

You will need

Serratia marcescens culture
Petri dishes containing nutrient agar of two different types
inoculating loops
Bunsen burner
access to an oven at 37 °C

Although you cannot easily use trees for an investigation into the contribution of inheritance and environment to variation because they would take too long to grow, there are other organisms you can use. Micro-organisms, that is organisms too small to be seen without the help of a microscope, generally grow and multiply rapidly. In this experiment you will use a bacterial micro-organism. Bacteria are usually very small (2×10^{-6} m is an average length).

Because many bacteria are able to cause sickness and illness (although those you are using in this experiment are normally harmless) a strict code of rules must be followed when experimenting with them. You will be shown the precautions and techniques necessary.

Note the appearance of the bacteria in the stock culture and then transfer some to each of two Petri dishes containing the nutrient jelly (agar) on which the bacteria feed. You will note that there are two sorts of nutrient agar. Incubate the dishes in an oven at 37 °C. After a few days note the appearance of the new colonies which have grown from those you originally transferred from the stock culture.

Discuss what should be done in the second phase of the experiment and then carry out your plan. How far do the results of this experiment overcome the objections referred to above?

How would you set about distinguishing between variations of an organism due to inheritance and variations due to the influence of the environment?

Investigation 4.11 The problem of the yellow plants

You will need

supply of one of the following types of seed: tobacco, barley or tomato
plant pots
seed compost

Most of you will be familiar with the fact that plants grown in the dark are a pale yellow colour instead of green. Grass which has been covered by an object for a period of time turns this yellow colour. Sow some seeds thinly in seed compost. Place them in total darkness. When they have started to grow examine their appearance and record it. Is it what you would expect?

Leave the seedlings for a few days in the light. What happens? ▶Explain your results.◀

Investigation 4.12 Making use of inherited variation

The pairs of photographs in figure 4.20 are obviously each of the same type of organism. Discuss how the variations they show may

Figurc 4.20

be of service to man. Discuss other examples in which man makes use of the variations in organisms.

Because a variation in an organism may be useful to man we often wish to exaggerate it in order to make it even more useful. This is done by selecting individuals in which the variation is more marked than in others. The selected individuals are allowed to breed together and the process is repeated on their offspring. The optional investigation which follows will give you a chance to perform a series of breeding experiments over a period of about one year. If a group of you decide to carry out this breeding programme remember that you will be taking on the responsibility of caring for living organisms for a long time and for keeping records.

▷ Investigation 4.13 Breeding from selected variations

You will need

pair of unrelated mice (male and female)
cages (about five)
supply of food

The variation you will study is in the mass of the mice. In each generation you will select the heaviest and lightest pair from which to breed. You should discuss with your teacher the overall strategy of the investigation and the techniques you will need to use.

In broad outline the procedure could be as follows:

Day 1	Mate the male and female mice.
Four weeks	Sex the offspring (your teacher will show you how to do this) and separate males and females.
Six weeks	Check the sexing and weigh all individuals. Select the heaviest pair for mating – this will be the basis of the 'heavy' breeding line. Select the lightest pair for mating – this will be the basis of the 'light' breeding line. Repeat the process.

You must keep careful records of matings, dates, masses of mice and so on. You can decide the way to do this by discussion amongst yourselves and with your teacher. You may also be asked to perform some experiments on the amount of food eaten by the mice.

5 Cells and more cells

In Section 4 you used a microscope to examine and measure the sizes of some very small organisms. We tend to take microscopes for granted but in the seventeenth century, when they were first developed, a whole new and unseen world was revealed. It is always exciting to look through a microscope so you can imagine the feelings of those who had the first opportunities to investigate what was for them an entirely new world! It is hardly surprising that they wished to look at living things. They wanted to know how organisms are built and especially if there were any patterns to be found in the way organisms are constructed.

Investigation 5.1 Investigating the structure of organisms with a microscope

You will need

fresh piece of onion
young leaf from *Elodea* plant
tooth-pick
microscope, slides and coverglass
dropper pipette
forceps
mounted needle
methylene blue solution OR
dilute iodine in potassium iodide solution
selection of prepared slides of parts of a variety of plants and animals

In this investigation you will look first at the structure of an onion, of Canadian pondweed - often grown in aquaria - and yourselves.

Separate a fleshy segment of onion from its neighbours. Snap the segment into two and see if this provides a small piece of ragged, almost transparent, 'skin' at the broken edge. Grip this skin gently with the forceps and peel off a larger segment. Place it in a drop of water on a glass slide (avoid folding the skin over on itself) and gently lower the coverglass over it with the mounted needle (figure 5.1). This will prevent air bubbles being trapped. (Air bubbles appear as black curved lines. If you do have some

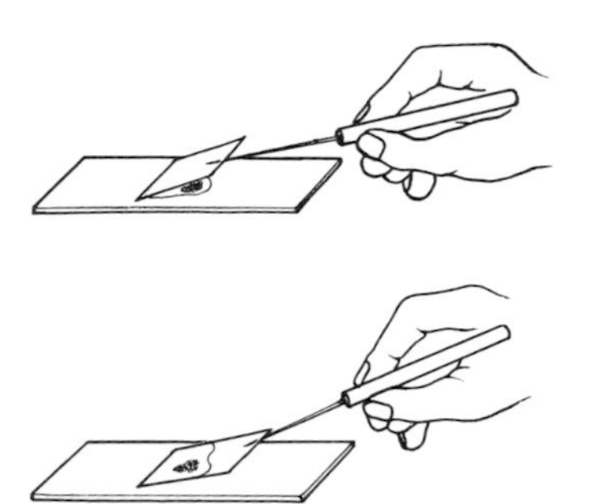

Figure 5.1
Lowering a coverglass on to a slide

trapped bubbles choose an area to look at that is clear of them.) Examine the slide under the low power and then the high power of the microscope. Describe what you see (sometimes a quickly made drawing is the best form of description).

Put the specimen on one side for use later. If the slide begins to dry, add more water to the edge of the coverglass with a dropper. Remove a young leaf from the tip of a Canadian pondweed plant and place it in a drop of water on a slide and add a coverglass. Examine it under the low power and high power of the microscope. Describe what you see.

Gently scrape the inside of your cheek with the broad end of a toothpick. Stir the scrapings into a drop of water on a clean slide. Add a coverglass and examine. Describe what you see. Put the specimen of cheek cells on one side for use in the next investigation. In your work so far can you begin to see a pattern about the structure of living things?

The building blocks of organisms you have observed are called cells. Measure the sizes of some of the cells you have observed.

Using each of your three specimens add some dilute iodine in potassium iodide solution or methylene blue solution to the edge of the coverglass. Draw it through by soaking up the water on the opposite side of the coverglass with a freshly torn corner of a piece of filter paper. Examine the slide with the microscope. Has the chemical stain made any difference?

How are cells constructed?

How do plant cells compare with animal cells?

Examine some prepared slides of parts of other plants and animals and from different parts of the same organism. These will probably have been stained to make the structure of the specimen clearer. You began your life as a single fertilised egg cell. You now consist of about 10^{13} to 10^{14} cells! How has this come about?

Is there a pattern in the structure of organisms?

Is there a pattern in the structure of cells?

Investigation 5.2 The structure of *Amoeba* and a mould

You will need

prepared microslides of *Amoeba* and mould
microscope

In this investigation you will examine a small animal known as *Amoeba* and part of a mould similar to the sort commonly found growing on bread. ▶Predict how they will be constructed. Examine them with a microscope.◀

Investigation 5.3 Structure and function in cells and tissues

You will need

colour transparencies of the specimens listed in the table below
slide projector

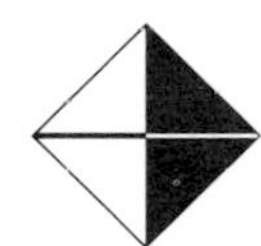

Examine colour transparencies of specimens of the cells and tissues listed in the table. Since tissues are made up from similar types of cell there is little point in considering tissues and cells separately when investigating their structure and the job they perform. However, a few types of cell can be considered in isolation and these are indicated in the table. As you study the cells and tissues look for a pattern.

cell or tissue	approximate size of cells	function (the job they do)
1 cheek	5×10^{-5} m across	lining layer of mouth
2 nerve	10^{-4} m across: fine fibres growing from cells may be 1 m long	carrying signals to and from parts of an organism
3 sperm cell	3×10^{-5} m long, including tail	swimming to and fertilising an egg cell
4 tendon	2×10^{-5} m	joins muscle to bone
5 xylem tissue	very varied: average size: 10^{-4} diameter; length may be well over 10^{-3} m	carrying water up the plant
6 outermost layer of cells on a leaf	3×10^{-5} m	protection
7 bone	10^{-5} m	support

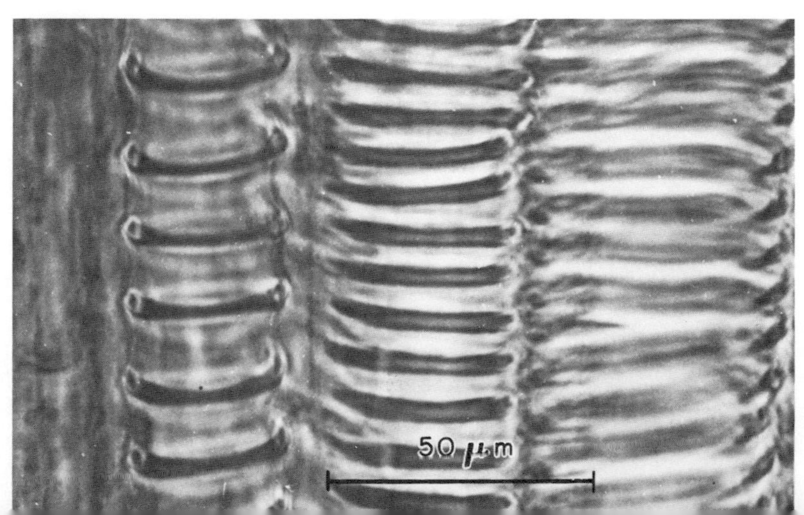

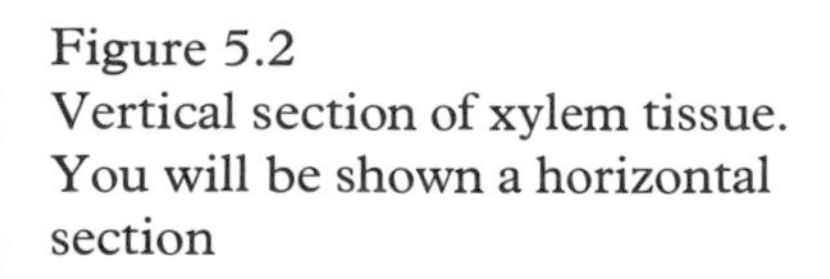
Figure 5.2
Vertical section of xylem tissue. You will be shown a horizontal section

Figure 5.3
Electron microscope photograph of cells lining the intestine where food is absorbed. How are they suited to perform this job?

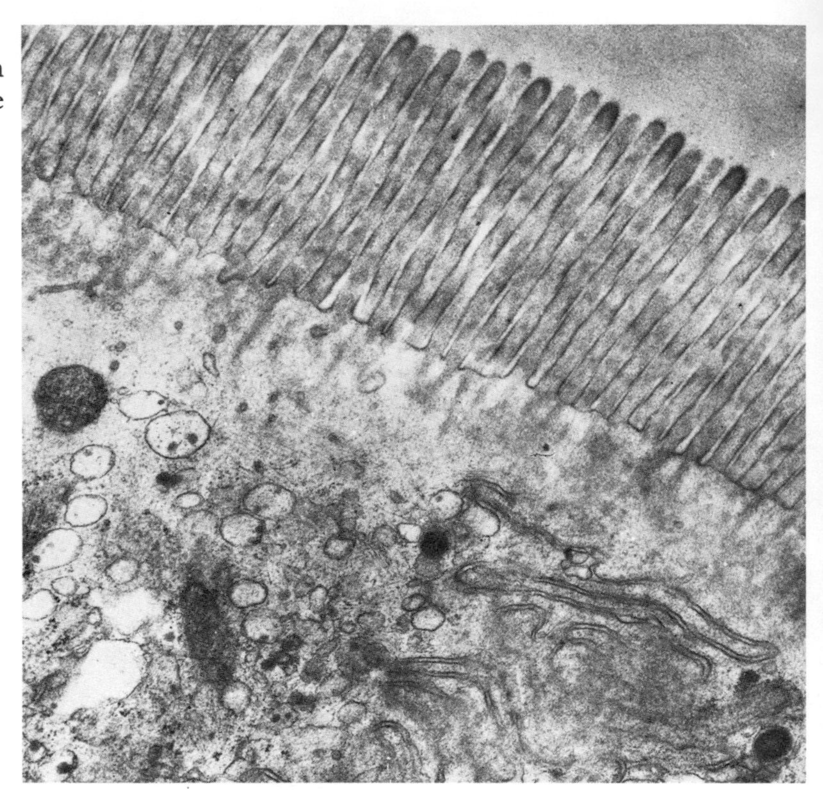

How has technology influenced the study of cells?

Investigation 5.4 Structure and function in an organ

You will need

horizontal section(s) of a plant stem
microscope

Study the pattern of arrangement of the tissues in a horizontal section of a plant stem. ▶From your knowledge of the pattern discovered in Investigation 5.3 explain how this organ (i.e. a structure made up from a collection of tissues) is suitably constructed to perform its function of supporting leaves and flowers.◀

▷Examine the second horizontal section of a plant stem which you may be given. From what sort of plant does this section come?

What parts of xylem cells are economically important and why?

6 Molecules

Molecules, atoms, ions and electrons are the smallest building blocks in *Patterns*. They are all closely connected, as you will see in the rest of Part 1.

In this section you will be answering questions such as 'How big are molecules?', 'What are the constituents of molecules?' and 'How can patterns of gas, liquid and solid behaviour be explained?' The first two investigations illustrate a particular property of gas molecules. Look for the common pattern in the experiments.

Investigation 6.1 Experiments with gases

You will need

two gas jars and covers (or two large test tubes)
lime water
supply of carbon dioxide
supply of town gas

If you invert an open gas jar full of a dense gas on top of a gas jar full of less dense gas, you might expect the dense gas to flow into the lower jar and displace the less dense gas. Try this with a jar of carbon dioxide molecules and air (which is a mixture of less dense molecules). How will you test whether there is any carbon dioxide in the lower jar?

Now repeat the experiment, but put the carbon dioxide in the lower jar. Leave it for five minutes, and then test to see if there is any carbon dioxide in the upper jar.

You could also do these experiments using air and town gas (also a mixture of molecules). How could you test for the presence of town gas? Would you expect town gas, when placed in the upper jar, to find its way into the lower jar? Does it?

Your teacher will repeat these experiments using a coloured gas such as nitrogen dioxide or bromine.

What is the common pattern in all of the experiments? In Section 1 you used a simple kinetic theory model. Use it again to explain the pattern in Investigation 6.1.

Investigation 6.2 Using the kinetic theory model

You will need

shallow baking tin
six marbles, about 1 cm diameter
one large marble

Place the six marbles in the tray. Keeping the tray flat on the bench, move it in an irregular manner as shown in Figure 6.1.

Figure 6.1

Describe the movement of the marbles. Do they move in different directions or the same direction? Compare their speeds. What happens when they hit the sides of the tray? What happens when they hit each other? Add a large marble and compare its movement with the movement of a small marble. What happens when a small marble and the large marble collide?

Compare the patterns of behaviour of the marbles with the patterns observed in Investigation 6.1.

Hydrogen molecules are much smaller than the molecules in air. Also, they move much more quickly. Use this information to explain your observations in the next experiment.

Investigation 6.3 Gas diffusion

You will need

the apparatus illustrated in figure 6.2
soap solution
hydrogen
retort stand, clamp and boss

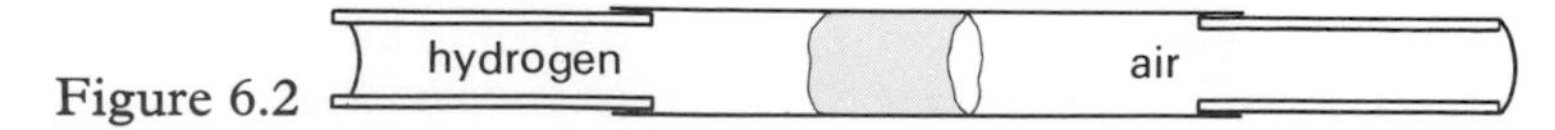

Figure 6.2

Gas molecules can pass (i.e. diffuse) through the chalk which is inserted in the polythene tubing.

Seal one end of the apparatus with a soap film. Fill the other end with hydrogen and seal with a soap film. Clamp the apparatus horizontally and watch what happens to the soap bubbles. ▶Explain your observations. Explain why the reverse happens with carbon dioxide.◀

Investigation 6.4 Brownian motion

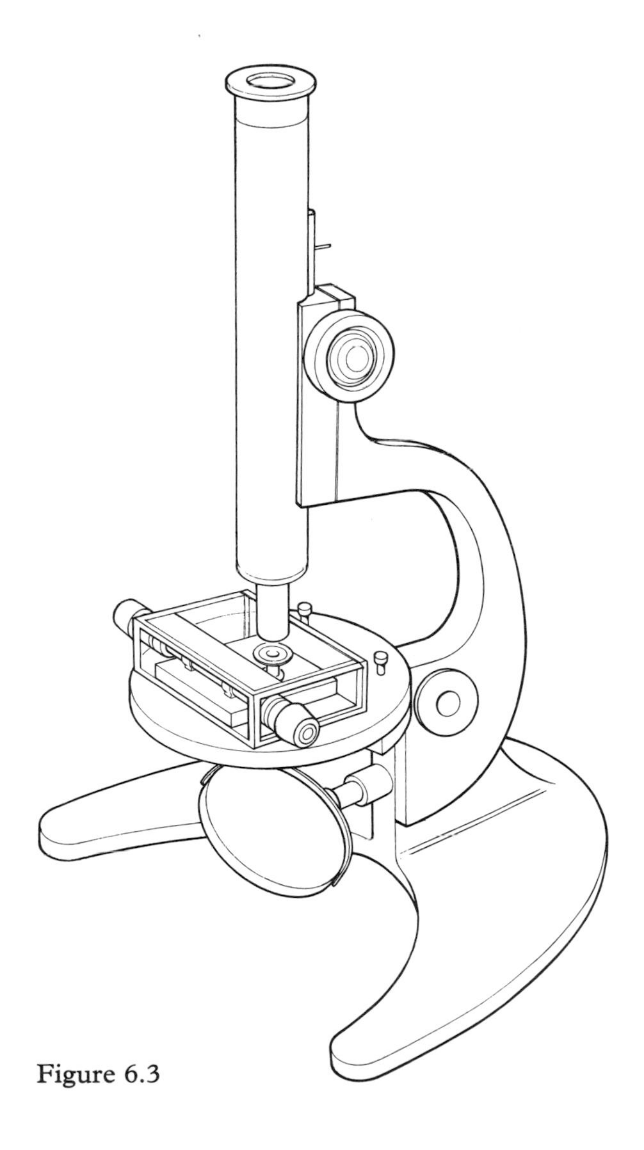

Figure 6.3

You will need

access to a microscope
Whitley Bay smoke cell (see figure 6.3)
dropper pipette

Remove the transparent plastic cover from the smoke cell and place the cell on the microscope stage. Connect the terminals on the smoke cell to a 12-volt electricity supply. Light the end of a piece of cord or straw and blow out the flame after a few seconds. Fill the teat pipette with smoke and then inject the smoke slowly into the glass cell. When it is full, seal it with the coverglass. Focus the microscope on to the top of the coverglass and then slowly lower the objective until you can see the smoke particles. Describe your observations.

▶Now use the kinetic theory model to explain your observations. Is the motion of the smoke particles in any way altered if the cell is warmed?◀

Investigation 6.5 Model of Brownian motion

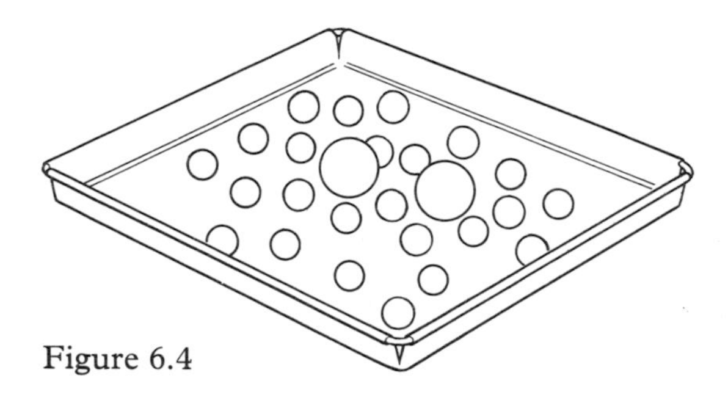

Figure 6.4

You will need

shallow baking tin
twenty-four marbles, about 1 cm diameter
two or three large marbles
empty matchbox

Repeat Investigation 6.2. First of all use the large marbles in the tray with the smaller ones (see figure 6.4), and then use the matchbox with the small marbles. Notice the effects of interactions of marbles with each other and with the matchbox. Describe what happens to the large marbles and the matchbox. How does this model help you to understand Brownian motion? The film loop *Movement of molecules* summarises much of the recent work.

Investigation 6.6 Making a prediction

▶ Using the kinetic theory of gases, predict what might happen to gas pressure when volume is decreased. Use the tin of marbles if you wish. Try to devise an experiment to test your prediction. ◀

Investigation 6.7 Compressing a gas

This is a demonstration to discover the effect of decreasing the volume of a gas on the pressure exerted by the gas.

The kinetic theory model used so far has had its limitations because it is flat (two-dimensional). A three-dimensional model might be of more use. Such a model is demonstrated in the next investigation.

Investigation 6.8 A three-dimensional model

Figure 6.5

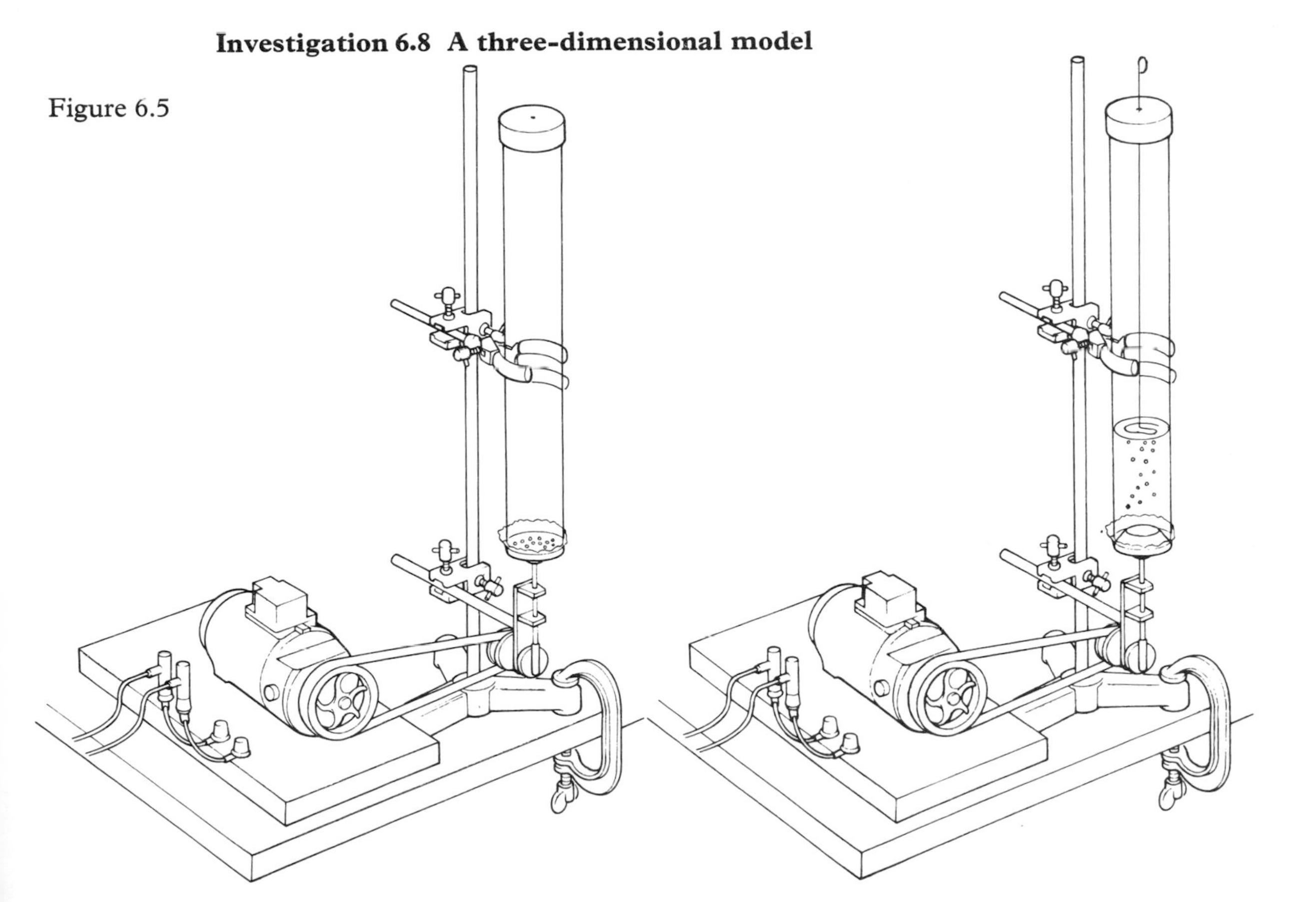

What happens to the pressure on the piston when the spheres move more quickly? ▶Is there anything which could be done to a real gas to make the molecules move more quickly?◀

Investigation 6.9 Warming the air inside a tin can

You will need

Bunsen burner
tripod
tin with push-in lid (*not* screw on lid)

Firmly close the tin and warm it over the Bunsen burner as shown in figure 6.6. What happens? ▶Explain your observations.◀ You could repeat this experiment with other gases (e.g. steam, carbon dioxide and nitrogen).

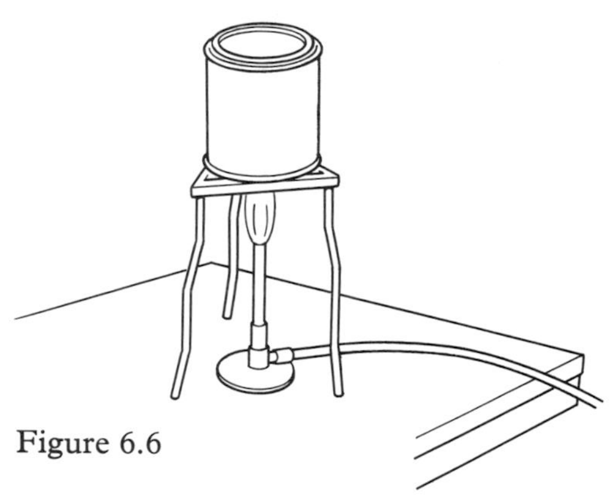

Figure 6.6

Why can it be dangerous to warm an unopened can of soup?
Summarise the patterns of gas, liquid and solid behaviour and the way these are explained by the kinetic theory model.
Is 'bouncing putty' a solid or a liquid?
So far you have been mainly concerned with the properties of gases, although the kinetic theory model was extended to include liquids and solids in Section 1. You will now be considering liquid molecules.

Investigation 6.10 Experiment with a coloured crystal

This is a demonstration experiment using iodine crystals and tetrachloromethane. The building blocks in iodine and tetrachloromethane are molecules. ▶Explain your observations using the idea of particles in motion. You could repeat Investigation 6.2 (but using 24 small marbles and tilting the tray slightly) to assist you in your explanation. Alternatively, you could just think about the experiment.◀

Although you cannot yet answer these questions you could start thinking about:
what keeps liquid molecules together?
what keeps solid molecules together?
what keeps gas molecules apart?

▷Investigation 6.11 Diffusion in liquids: osmosis

Part a

You will need

solution of copper sulphate 0.25M
beaker, 100 cm^3
crystal of potassium hexacyanoferrate (the size of a pea)

Pour the solution into the beaker. Add to the solution the crystal of potassium hexacyanoferrate. Watch the crystal. Describe what happens.

Part b

You will need

solution of sodium silicate (water glass: 30 g of sodium silicate in 130 cm^3 of water)
beaker, 250 cm^3
crystals of iron (II) sulphate, copper sulphate, cobalt sulphate and nickel sulphate

Drop the crystals into the solution of sodium silicate and leave overnight. Describe the 'chemical garden' which is formed by the next day.

Part c

This is a demonstration experiment.

What is the pattern common to all three parts of this investigation? In what ways does this pattern compare with the observations made in Investigation 6.3?

This pattern is called osmosis. Osmosis is important to living systems as the next investigations show.

▷Investigation 6.12 Plant cells and solutions

The purpose is to investigate the effect of placing living plant cells in

a tap water (a dilute solution)
b strong sucrose solution (a concentrated solution).

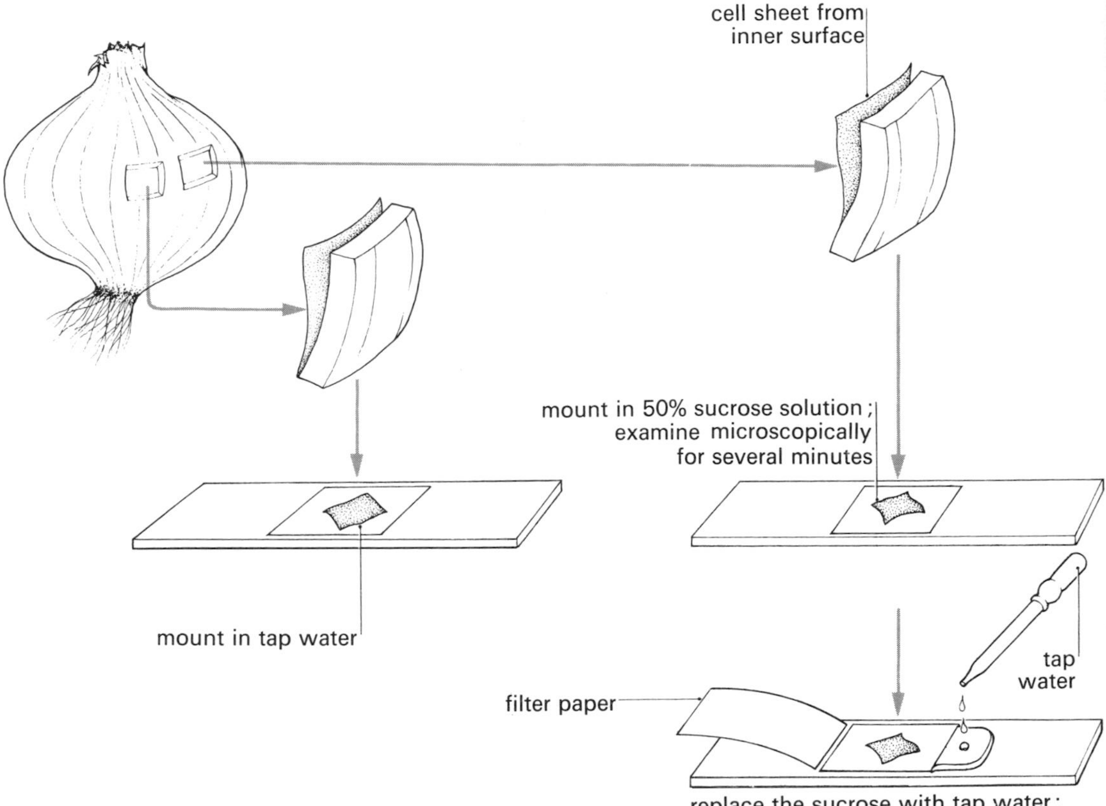

Figure 6.7 Preparing and irrigating sheets of onion cells

You will need

onion (preferably one of the red varieties) or rhubarb
microslides and 2 or 3 coverglasses
small quantity of 50% sucrose solution in a beaker
dropper pipette
several strips of filter paper for irrigation
forceps
camel-hair brush
microscope

This method is shown in the flow diagram in figure 6.7. The onion bulb is a good source of living cells since sheets one cell thick can be peeled from the inside of the bulb scales. Use a red-skinned onion if possible as the contents of the cell are coloured and thus easier to observe, but an ordinary white onion will do well enough.

Note the following precautions:

1 Do not let the sheet of cells dry out, or they will die. Therefore, transfer them to the solution quickly.

2 In using forceps to peel off the cells remember that any cells gripped in the blades will be damaged. Therefore, use a brush to manipulate the cells once they are removed.

Sketch the appearance of the cells under the microscope at the various stages. What is the 'variable' in the experiment? Can you think of other variables? Which sheet of cells is the control? Figure 6.8 will remind you of the structural pattern of a plant cell. Somc of the smaller details have been omitted. The vacuole contains a solution consisting of water with various substances dissolved in it.

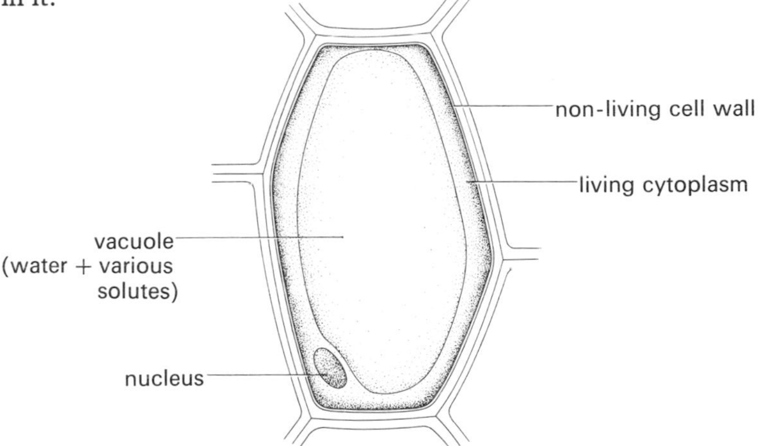

Figure 6.8
A simplified generalised plant cell in two dimensions

In your experiment did the plant cell wall expand away from the rest of the cell or did the rest of the cell shrink from the cell wall? What happened to the inside of the cell in the strong sucrose solution? What happened to it when it is replaced in tap water (the dilute solution)? ►Can you explain these observations? The use of the cell models may help to develop your ideas.◄

▷Investigation 6.13 Using cell models

This is a demonstration experiment.

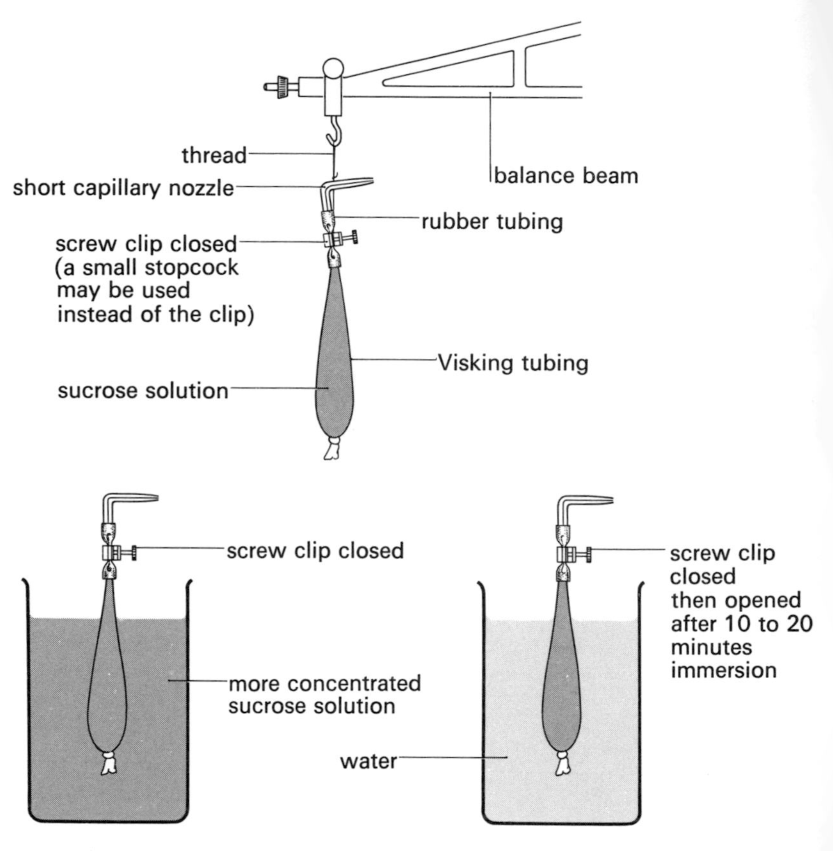

Figure 6.9

The previous investigation may have made you suspect that the cytoplasmic lining and vacuole are responsible for the changes you observed. We can build a model of the cytoplasmic lining and vacuole using a material known as Visking tubing and you will see this demonstrated. Figure 6.9 shows you the sequence of events. What happened to the mass of the 'cell' in air? Explain your observations.

What happened to the 'cell' in the more concentrated sucrose solution? What happened to the 'cell' in tap water? Did sucrose pass through the membrane?

Do these observations provide a basis for formulating a hypothesis about what actually happens in real cells as observed in the previous investigation? What could be the partially (or semi-) permeable membrane in cells?

The process of removing water from, and then adding water to, food cells is an important factor in food preparation. In order to prevent decay and thus preserve food, the water contained in the cells must either be removed or solidified. If it is removed it must later be replaced. The earliest ways of removal were to expose the food to the sun or to warm it in front of a fire. These methods of removing water (called dehydrating) are still used (for example, in curing fish or making dried peas).

Clarence Birdseye first developed the technique of 'quick-freezing' on a commercial scale. In this process ice crystals are formed rapidly between and inside the cells. The crystals therefore remain small and so do not damage the cells.

Figure 6.10
Peas passing through a tunnel used in quick freezing

Discuss the figures shown in the following table. What do they reveal about the popularity of quick frozen foods?

UK consumers' expenditure on food 1959–1969

	food/£million	quick frozen food/£million
1956	4 274	16.0
1957	4 448	22.5
1958	4 493	31.5
1959	4 647	42.0
1960	4 723	53.0
1961	4 885	66.0
1962	5 112	72.0
1963	5 262	75.0
1964	5 495	80.75
1965	5 691	90.25
1966	5 972	108.5
1967	6 152	122.25
1968	6 358	141.5
1969	6 705	157.0

Accelerated freeze-drying is a new process where the food is first quick-frozen and then subjected to a vacuum. The ice crystals turn to vapour immediately, leaving the food almost completely dry.

▷ Investigation 6.14 Comparing preserved foods

You will need

dried peas
Batchelor's 'Surprise' peas
frozen peas
canned 'garden' peas
fresh peas (if obtainable)

Note the mass, space occupied, texture and appearance of the foods.

Prepare, cook and eat the foods following instructions from your teacher, or those on the packet or can. Find the preparation time before cooking. Find the time needed to cook the food.

There are a number of important features which we, as consumers, are concerned about in respect of the food we eat. They include relative costs, appearance, taste, ease of preparation and ease of storage. Make an assessment of these features for the foods you have used. Can you suggest possible reasons for the trends shown in the following table?

how we buy peas	percentages of peas bought						
	1963	1964	1965	1966	1967	1968	1969
quick-frozen	26	28	29	30	33	34	34
canned garden	23	21	20	20	19	18	18
canned processed	25	25	25	25	25	25	25
others (fresh, dehydrated etc.)	26	26	26	25	23	23	23

Assess the advantages and disadvantages of different forms of food preservation, comparing them with each other and with the fresh food.

Think about the following:
transport from harvesting to factory
transport from factory to shops and consumers
storage of preserved food in warehouses, shops and homes

ease of preparation and
consumer 'acceptability' (e.g. relative costs, appearance and taste, presence of preservatives and artificial colouring).

The last two items give you an opportunity for some experimental work and surveys which you might like to try in your spare time at home and school.

Already there have been hints in this section that molecules vary in size. In the next investigation you will get some idea of the upper limit for their size.

Investigation 6.15 How big are molecules?

You will need

flat dish or tray
black wax
two wires mounted on cards
Lycopodium powder
two metal booms
olive oil

You have been shown the purpose of the wax edges to the tray together with how to set up the tray and level it full of water.

The following list of instructions will remind you of the many things you must do.

Cleanliness is essential.

1 Wash your hands before starting work and try not to touch any part of the tray or other apparatus which will come into contact with the water.
2 Inspect the edges of the tray and the metal booms. Look to see if there is a nearly uniform layer of black wax. If not, effect a quick repair using the hot wax available.
3 Position the drain hole over the sink. Level the tray using four rubber wedges. Fill the tray with water and adjust the level until the water surface is proud all the way round. Draw the booms across the water surface to clean it of any dust.
4 Prepare a single drop of oil on the fine wire mounted on the card. Inspect the drop using the lens and transparent glass scale, teasing surplus oil into the drop or removing it, by using the other card-mounted wire. Estimate the diameter of your drop.
5 Lightly sprinkle the surface of the water with the *Lycopodium* powder (which floats and does not absorb water).

6 Place the drop of oil, which is on the wire, in the centre of the tray by dipping the wire loop below the surface of the water. The oil film should spread over a large area, pushing the *Lycopodium* in front of it. If this does not happen the surface has been contaminated and should be cleansed with a boom. After the circle has finished expanding and before it distorts, measure its diameter quickly.
7 Carefully draw a boom over the surface pushing the oil and *Lycopodium* to one side. Place the second boom behind it on the clean side of the tray and rinse the one you have used.
8 Repeat the process again to obtain a second set of results.

There are two ways of calculating the size of the oil molecule. Both are given.

Method 1

Assume the drop to be a cube of side length equal to the diameter you measured. What is the side length in mm? What is the side length in m? (Use power of 10.) Volume of cube = (length of one side)^3. What is the volume of your oil in m^3?

Assume the circle to be a square plate of oil of side length equal to the diameter you measured. The size of an oil molecule will be approximately equal to the thickness of the oil film. (What assumption is being made here?) If d is the thickness of the plate you will need to find d.

Why do you need to know d? Volume of disc = $(\text{length})^2 \times d$. What was the length (diameter) of your plate of oil in m? Where did the oil come from? What has the same volume as the plate of oil? Calculate d in metres.

Method 2

Work always in metres. Volume of sphere $= \frac{4}{3}\pi r^3$ (remember, you measured the diameter of the drop). Calculate the volume in metres.

The circular oil film has volume $\pi R^2 d$, where R is its radius and d its thickness (remember, you measured its diameter). Calculate the area of the drop in metres. Volume of oil drop = (Area of oil film) $\times$ d. Calculate d in metres.

Is this method of calculating molecular size likely to be accurate?

There is a considerable variation in the sizes of molecules. This can affect the properties of the molecules. Polyethylene is so large that it is termed a 'giant molecule'.

▷Investigation 6.16 Bending polyethylene

You will need

strip of low-density polyethylene
strip of high-density polyethylene

Low-density polyethylene consists of fairly short molecules and high-density polyethylene consists of long molecules. ▶Guess which of the strips you would expect to be the more flexible, explaining why you made that decision. Bend both the strips in turn. Identify the strips. Check your answer with your teacher.◀

▶Devise an experiment to make more precise measurements and then attempt the experiment.◀

Investigation 6.17 Some physical properties of alkanes

Alkanes, which are molecules of different lengths, are made from similar units which are joined together as chains. In the following table these units are shown as 0. (Polyethylene giant molecules consist of about 2 000 of these units.)

name of alkane	chain	density/g $m^{-3} \times 10^5$	melting point/K	boiling point/K
methane	0	4.2	90.7	111.7
ethane	0–0	5.5	89.9	184.5
propane	0–0–0	5.8	85.5	231.1
butane	0–0–0–0	5.8	134.8	272.7
pentane	0–0–0–0–0	6.3	143.4	309.2
hexane	0–0–0–0–0–0	6.6	177.8	341.9
heptane	0–0–0–0–0–0–0	6.8	182.5	371.6
octane	0–0–0–0–0–0–0–0	7.0	216.4	398.8

Describe (in general terms) how the densities, melting points and boiling points of the alkanes alter with increased molecular size. ▶Give an explanation for this pattern based on the kinetic theory model for liquids and gases.◀

Molecules of different lengths (which are rather like alkanes) are mixed together in crude oil. The pattern you have just found is used in purifying the oil. Find out more about this from reference books.

Investigation 6.18 Separating molecules of different sizes

You will need

retort stand, clamp and boss-head
Bunsen burner and hardboard mat
hard glass test tube with side-arm, 125 × 16 mm
bent delivery tube and rubber connecting tube
four test tubes, 75 × 12 mm
thermometer (0 ° to 360 °C) and cork
dropper pipette
beaker, 100 cm^3
watch glass (hard glass)
Rocksil
crude oil, about 2 cm^3

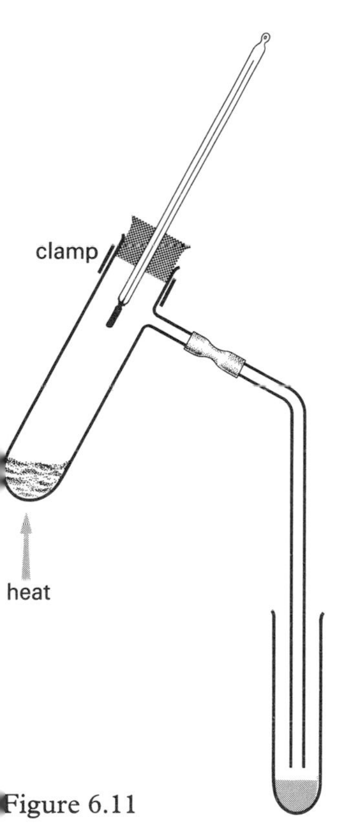

Figure 6.11

Place a loose plug of Rocksil in the bottom of the side-arm test tube and add the crude oil, using the dropper pipette. After setting up the apparatus, warm the test tube carefully.

Use the four test tubes to collect the oil which distils within the following temperature ranges:

test tube 1: room temperature to 70 °C
test tube 2: 70 ° to 120 °C
test tube 3: 120 ° to 170 °C
test tube 4: 170 ° to 220 °C

▶Which of the test tubes contains the shortest molecules and which contains the longest?◀ How viscous (i.e. how 'runny') are the samples? What is the pattern of viscosity? Describe the colours of the samples.

Do they burn easily? Pour each sample in turn on to a small piece of Rocksil on the watch glass and light with a splint. Compare the flames.

This process of purification is called 'fractional distillation' and each of the samples of purer oil you collected is called a 'fraction'.

Discuss some of the effects the production of pure oils has had on our standard of living.

Molecules can form crystals (e.g. sugar) and in a later investigation you will be watching molecular crystals of naphthalene grow.

Investigation 6.19 Dialysis: a further consequence of building block size

Thomas Graham (1805–1869) discovered that some substances when mixed with a liquid will pass through a membrane, whereas others will not.

You will need

dialyser and membrane
crystallising basin
0.1M sodium chloride solution; 1% starch solution
silver nitrate solution; iodine/potassium iodide solution

Set up the apparatus as shown in figure 6.12. As soon as the experiment is started test a few drops of the surrounding liquid for starch and for the chloride building block. Repeat the tests after 30 minutes.

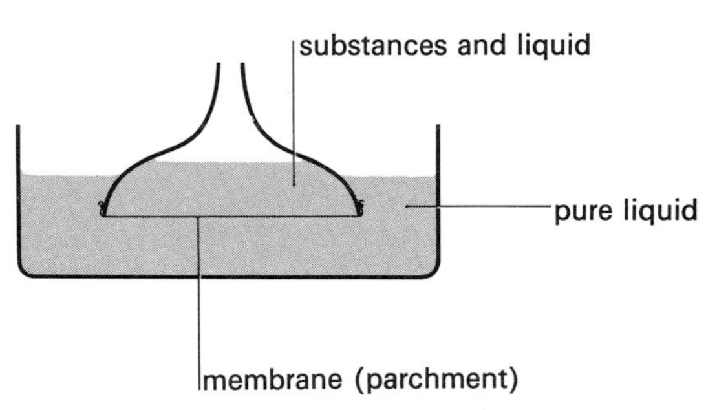

Figure 6.12

The results of similar experiments for other substances are given:

substance	liquid	A	B
sodium chloride	water	small	
starch	water	large	
gum	water	large	no
sodium chloride	benzene	large	no
silver nitrate	water	small	yes
soap	water	large	no
sugar	water	small	yes
protein	water	large	no

Column A gives a description of the building block size when mixed with liquid and column B states whether the building block passed through the membrane.

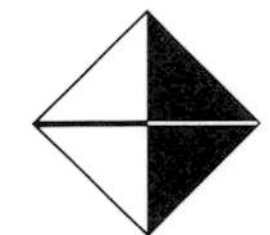

What is the pattern linking size of molecule (or other building block) and passage through the membrane? Offer an explanation for this pattern.

Investigation 6.20 How does food get into the body?

When food has been chewed and swallowed it passes through the body in a tube called the alimentary canal (or gut). It is a single, continuous tube running from the mouth to the anus. Examine the drawing in figure 6.13. If you consider the area labelled A which represents the inside of the gut you could describe any food present as within, but still outside, the body. ►How, then, does food get through the wall of the gut into the body? In this investigation we shall attempt to answer this question.◄

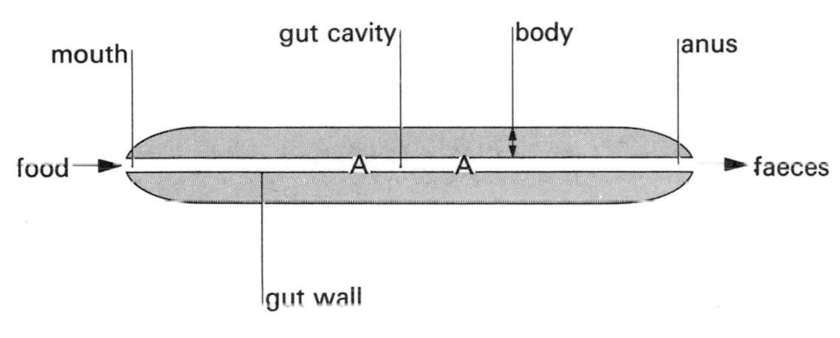

Figure 6.13
A simplified plan showing the relationship of the gut to the rest of the body

You will need

iodine/potassium iodide solution to test for starch
Benedict's solution to test for sugars
1% starch solution (Analar grade) containing 0.1% sodium chloride, 30 cm^3
Visking (cellulose) tubing, 14 mm diameter – three pieces about 150 cm long
thread
wax pencil
saliva
paper clips
plastic syringe, 10 cm^3
beaker, 400 cm^3
three test tubes, 125 × 16 mm
three test tubes, 75 × 12 mm
thermometer, –10 ° to 110 °C

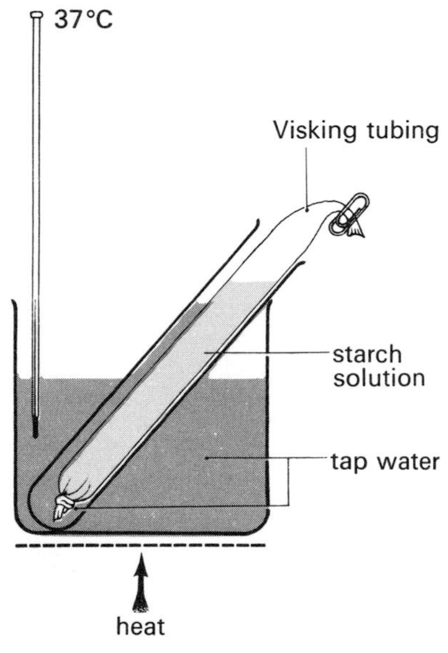

Figure 6.14

Bunsen burner, tripod and gauze
dropper pipette
spotting tile

Mix 10 cm^3 of starch with one-third of the saliva in a clean beaker. Firmly tie one end of each piece of Visking tubing with thread to make it into a bag. (The Visking tubing can be handled more easily if it is wet. One end can be opened by rubbing between thumb and forefinger.)

Fill the three 'bags' as follows, using the syringe:
a starch solution only
b saliva solution only
c starch and saliva mixture

Wash each bag thoroughly in running water to remove any trace of starch and saliva from the outside of the cellulose bag. Do not let the water get inside. Label each test tube before you put it in the beaker and set up the apparatus as shown in figure 6.14. You will have three test-tubes in the beaker. Immediately withdraw a sample of water from each test tube and test it for starch and sugars. After 15 minutes repeat the tests. If they are still negative wait a little longer and then repeat the tests. ▶Record your results carefully and explain them fully. As a result of this explanation suggest how food gets into the body.◀

Investigation 6.21 The digestive system

▶This is a demonstration investigation.◀

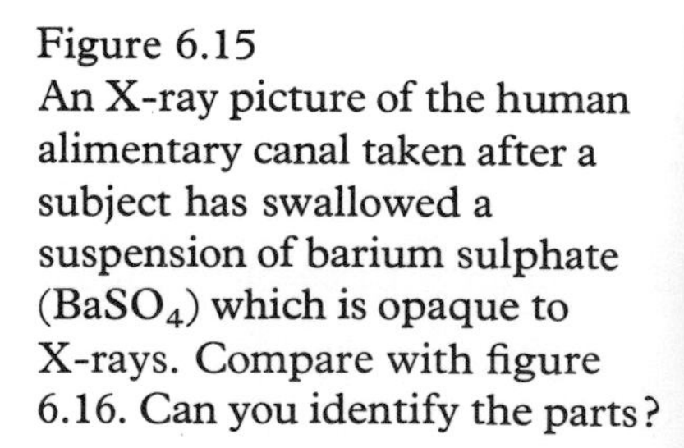
Figure 6.15
An X-ray picture of the human alimentary canal taken after a subject has swallowed a suspension of barium sulphate ($BaSO_4$) which is opaque to X-rays. Compare with figure 6.16. Can you identify the parts?

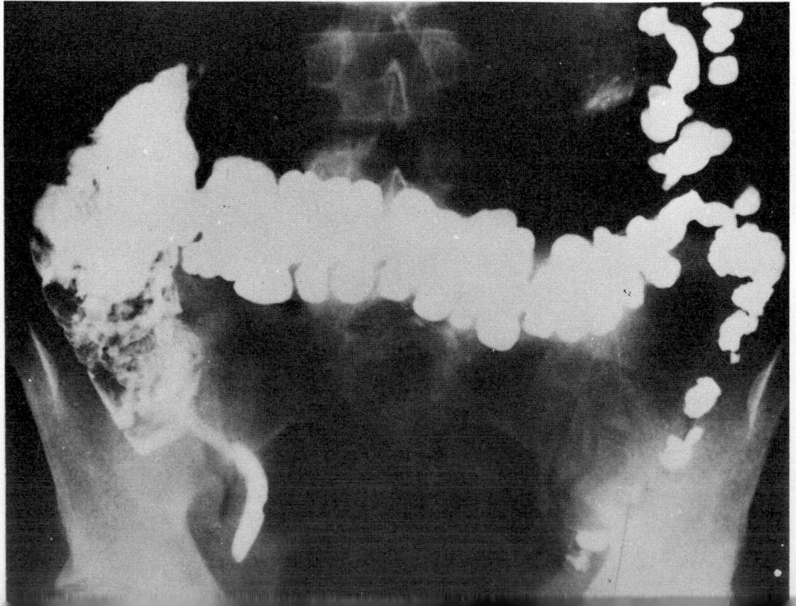

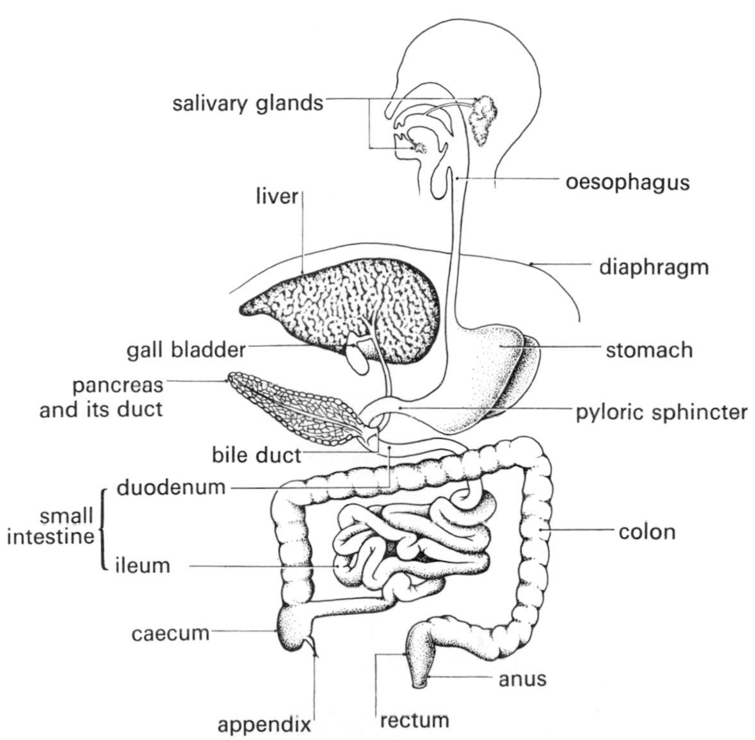

Figure 6.16
A diagram of the human digestive system. The chest has been omitted, and the small intestine considerably shortened

Figure 6.17
A view of the inside of the small intestine where food is absorbed. The finger-like projections extend into the cavity of the gut and are called villi. How might they assist the absorption of food?

Investigation 6.22 The artificial kidney

The artificial kidney works in approximately the same way as the real kidney although its action is less complicated. Unwanted substances (other than water) are removed from the blood by a process of dialysis.

The patient's blood flows on one side of the Visking membrane (figure 6.18a shows it being prepared) and a dialysing fluid on the other. Smaller building blocks diffuse out of the blood and are carried away as waste.

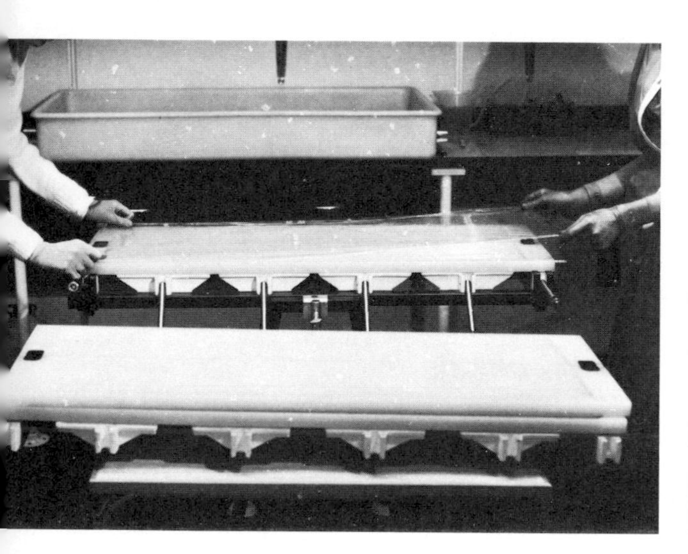

Figure 6.18a
Preparing the Visking membrane

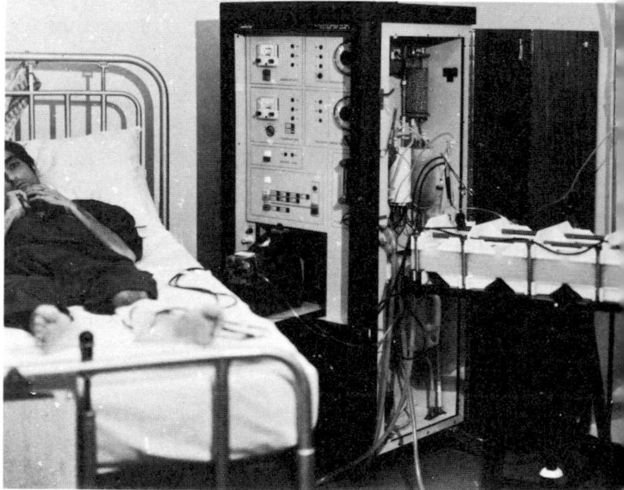

Figure 6.18b
The kidney machine in action

Excess water is removed from the body by increasing the patient's blood pressure, causing water to be forced through the Visking tubing.

There are too few artificial kidney machines. A hospital may have to refuse some patients. Who should make these decisions? What facts do you think they should consider before deciding? Has science any contribution to make in these decisions?

▷ Investigation 6.23 Molecular crystals: naphthalene

You will need

two test tubes, 100 × 16 mm

dropper pipette
spatula
microscope slide
microscope
tetrachloromethane
naphthalene

Add one small measure of naphthalene to each of the test tubes. One-quarter fill one of the test tubes with water and to the other add a similar volume of tetrachloromethane. Shake the test tubes. In which liquid does naphthalene dissolve? Use the dropper pipette to put one drop of this solution on to the microscope slide and watch what happens.

Investigation 6.24 Molecular crystals: sulphur

You will need

pestle and mortar
beaker, 250 cm^3
watch glass
test tube, 100 × 16 mm
filter paper
hand lens or microscope
xylene
roll sulphur
small conical flask

Half fill the test tube with small pieces of roll sulphur. Fold a filter paper as though you were going to put it into a funnel, but instead fasten it with a paper clip and hold it with tongs. Warm the sulphur gently until it has just melted and then pour it into the filter paper. As soon as a crust has formed on top of the sulphur, remove the paper clip and open up the paper. Be careful not to touch the molten sulphur still inside. Do you see any crystals inside? If so, what shape are they? Make a drawing of them. These are crystals of monoclinic sulphur.

Turn out all Bunsen flames. Crush a piece of roll sulphur about the size of a rice grain, and shake it in about 25 cm^3 of xylene in a small conical flask. The xylene may be warmed in half a beaker of hot water. When most of the sulphur has dissolved, pour the solution onto the watch glass which is in the fume cupboard. Cover the watch glass with a piece of paper. After doing the next experiment, examine the crystals which have formed with a hand lens or microscope. Draw the crystals. Is their shape similar to those which were formed in the last experiment? These are crystals of rhombic sulphur.

Discussion with your teacher will show that the internal arrangement of molecules can explain crystal shape. Sugar is crystalline when viewed under a microscope and yet icing sugar is not. Does this mean that icing sugar has no internal structure?

Discussion with your teacher will have shown that molecule building blocks are made up from atom building blocks. For example, the sulphur molecule is made from eight sulphur atoms. Sulphur is called an element because its molecules are made from the same kind of atom building blocks. The alkane molecules in Investigation 6.17 are made from two different atom building blocks: carbon and hydrogen. Alkanes are examples of compounds because they are made of more than one type of atom.

One of the commonest compounds is water. In the next investigation its constituent atom building blocks will be separated.

Investigation 6.25 Breaking up water

Using an electrolysis cell and the acidified water provided, discover what happens when acidified water is electrolysed. Identify the gases evolved at the positive and the negative electrodes, i.e., at the anode and cathode.

Investigation 6.26 Making water from its elements

This is a demonstration experiment.

Three questions were raised at the start of the section. You should now know enough to answer the questions. One of the questions (What are the constituents of molecules?) will be answered in greater depth in the next section.

7 Atoms and giant structures

In the last section you learnt that there are molecules of different sizes and that these molecules consist of even smaller particles called atoms. 'Are there patterns of combination of atoms?', 'How big are atoms?' and 'How are atoms arranged in metals?' are just three of the questions to be answered in this section.

'Atom' is a word which all of us use in everyday conversation. We talk about atomic power, atomising Martians, the atomic bomb and so on. It may surprise you, therefore, to learn that scientists have never directly seen an atom; it is just a very useful idea for explaining a large number of observations.

The following experiments introduce some of the evidence for the existence of the atom as well as some of the uses to which the idea may be put.

Investigation 7.1 Seeing patterns indirectly

You will need

set of Nuffield diffraction grids
small light source (e.g. a torch bulb)
access to a low power microscope

The diffraction grids consist of white dots on a black background. The patterns of dots are either squares, or diamonds, or a random 'jumble'. Use the marked cards to show that the light pattern you can see is related to the actual pattern of white dots on the cards. To do this you should hold the card close to one eye; the other eye should be used to look through the grid to the light source. Now identify the patterns of dots on the unmarked cards. Check your answer with a low powered microscope.

This indirect way of identifying patterns of dots is also used to identify atoms in X-ray analysis and in electron microscopy.

Investigation 7.2 Looking at electron micrographs

The following electron micrographs have been obtained by passing

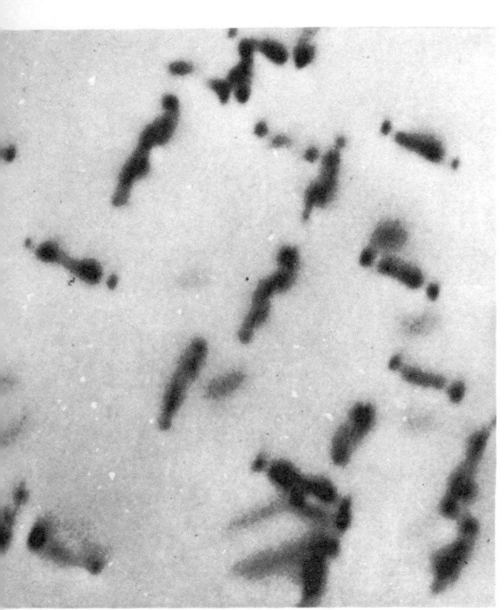

Figure 7.1a
Electron micrograph of diamond

electrons through samples and taking photographs of the patterns which are produced. The technique, which is similar to the previous investigation, need not trouble us here, but what happens is that an 'electronic eye' (the electron microscope) enables us to 'see' the structures of substances. What is 'seen' must now be interpreted.

Three scientists at the University of Chicago obtained the first photographs of individual atoms of uranium and thorium by using a newly developed electron microscope that magnifies up to 5 million times. (Do you find this fact exciting?)

The work at the University was headed by Dr Albert Crewe, a British-born physicist who was once the director of the Atomic Energy Commission's Argonne National Laboratory. Dr Crewe designed and built the electron microscope which made the feat possible.

The three scientists calculated that atoms of heavy elements would show up as bright spots against a dark background. This calculation was correct, as figure 7.1b shows, where two uranium atoms appear as relatively bright spots (arrowed). The white dots also showed up on the picture in figure 7.1c which shows Dr Crewe pointing out a single chain of thorium atoms.

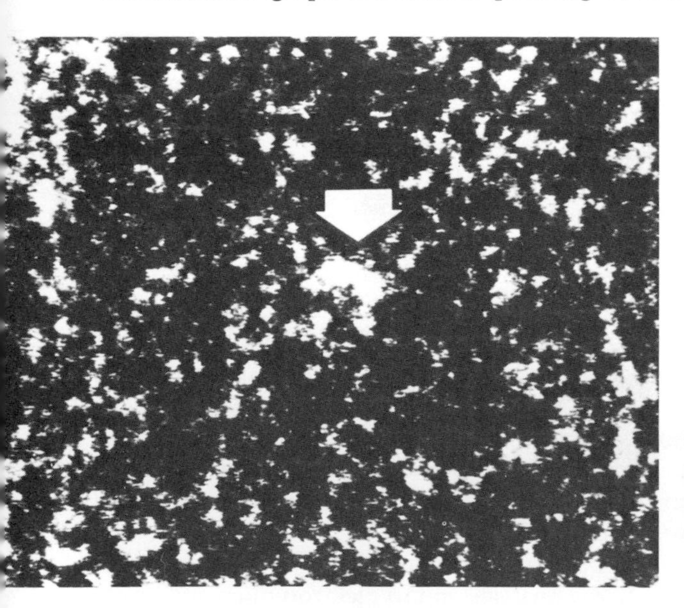

Figure 7.1b
Printed to a magnification of about $\times 25$ million

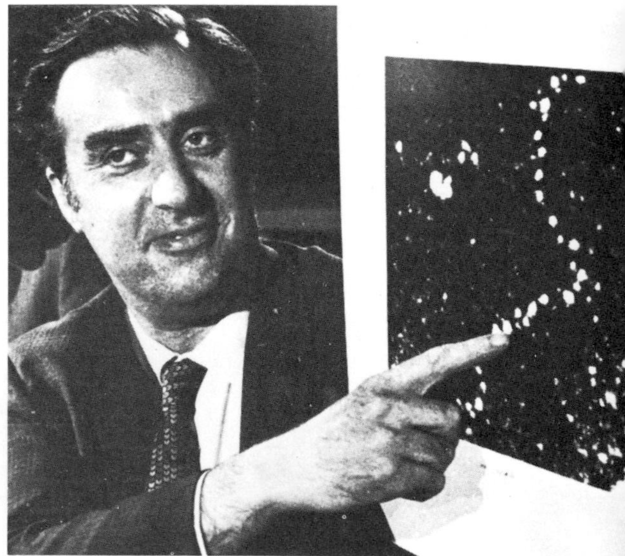

Figure 7.1c

What are the sizes of the atoms shown in the micrographs? How accurate are the measurements likely to be? (You could start by comparing the different answers obtained by your class.)

Investigation 7.3 Spectroscopic evidence for the atom

You will need

direct vision spectroscope rigidly clamped in position
piece of platinum or nichrome wire attached to a glass rod
test tube, 100 × 16 mm
filter funnel and filter paper
watch glass
evaporating basin
Bunsen burner, tripod and gauze
crucible tongs
sodium, calcium, potassium and strontium chlorides
copper oxide (black)
piece of copper about 2 × 2 cm
dilute hydrochloric, nitric and sulphuric acids
concentrated hydrochloric acid

Look through the spectroscope at the light from the window. What can you see? Clean the platinum wire by dipping it into concentrated hydrochloric acid and then warming the tip of the wire in a colourless Bunsen burner flame. Dip the tip of the wire into one of the chlorides contained in the watch glass and then hold the wire in the flame. Look at the flame first directly and then through the spectroscope. The spectroscope spreads the light into a spectrum. How does it compare with the spectrum you obtained from white light?

Repeat the experiment (remembering to clean the platinum wire each time) for the other chlorides. Are the spectra obtained from the chlorides similar or are they different from each other?

Hold the piece of copper in the flame using the crucible tongs. Look at the flame directly and then through the spectroscope. Comment on what you observe.

Groups in the class can make samples of copper chloride, copper sulphate and copper nitrate in the following way. To about 5 cm^3 of dilute acid add one measure of black copper oxide and warm the mixture. Cool and filter off the excess copper oxide, collecting the filtrate in an evaporating basin. Gently warm the evaporating basin until most of the water has boiled away, and remove the Bunsen burner. The heat of the gauze and tripod will be sufficient to remove the rest of the water, leaving the salt.

Repeat the spectroscopic analysis for each of the three salts which are formed (you will have to obtain two of the salts from other groups). How do the spectra for the copper salts compare?

How do the spectra compare with the spectrum obtained from pure copper? ▶What pattern emerges from the spectra experiments which you have done? Can you offer an explanation for this?◀

It would have been simpler in this investigation just to have taken samples of copper chloride, copper sulphate and copper nitrate straight from bottles. ▶Why would these samples not have provided evidence for the existence of the atom?◀

The Jackdaw *The discovery of the galaxies* shows one use of spectroscopy in identifying elements on other planets.

In the previous section you learnt that atoms can interact (or combine) to give molecules of various sizes. They can also interact in other ways; the next investigations show one of these ways.

Investigation 7.4 Another look at sulphur

You will need

pestle and mortar
beaker, 250 cm^3
test tube, 100 × 16 mm
test tube holder
universal indicator paper
roll sulphur

Figure 7.2
What is likely to happen to the sulphur molecule when sulphur is melted?

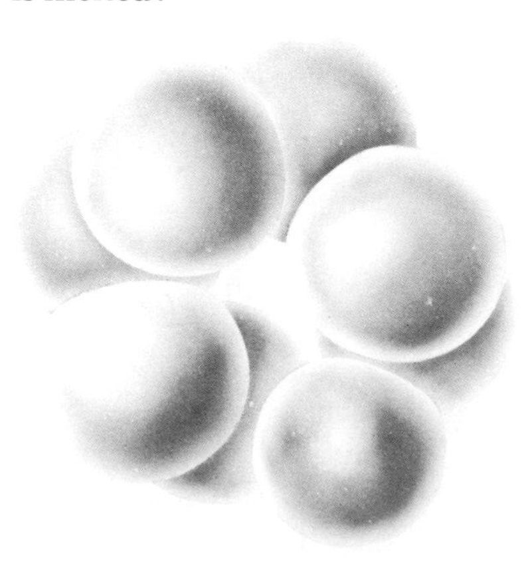

Ensure that the laboratory is well ventilated for the experiment. Write a short account of what you would expect to happen when a solid like sulphur is warmed to its boiling point. Half fill a dry test tube with small pieces of roll sulphur. Warm the sulphur gently with a low Bunsen burner flame. Describe (a) the colour and (b) the viscosity ('runniness') of the sulphur as its temperature increases. When the sulphur is boiling ignite the vapour. Smell the resulting gas and test it with moist universal indicator paper.

Pour the boiling sulphur into a beaker half filled with cold water. As soon as it is cool enough to touch, pick out the sulphur from the water and notice its properties. What special property has it? Examine this sulphur again after about 24 hours. Has it changed?

Compare your account of what actually happened when sulphur is warmed with your account of what you expected to happen. ▶If the assumption is made that plastic sulphur consists of thousands of atoms joined together in chains, explain the elastic property of the sulphur.◀ In fact, a chain can consist of a million sulphur atoms. This structure is obviously a giant of a giant 'molecule'!

▶What do you suggest must have happened to the internal arrangement of atoms in the plastic sulphur when it was allowed to stand for 24 hours?◀

Investigation 7.5 Giant structures

You will need

fourteen polystyrene spheres
cocktail sticks
pipe cleaners
cardboard protractor

One of the most useful ideas of modern chemistry is that the arrangement of the building blocks of which a chemical is composed plays an important part in determining the properties of the substance.

You will be familiar with non-chemical examples of the effect of making different arrangements of non-scientific building blocks. Here are some illustrations:

a If the basic blocks are integers, their arrangement in a number determines its value. 123 can be rearranged to give 132, 231, 213, 312 and 321. How many arrangements of 1234 are possible?

b Another example of a one-dimensional linear arrangement is the anagram – the rearrangement of the position of the letters in a word to form new words, e.g. heat . . . hate, tear . . . rate. Rearrange the letters in 'stop' to form as many genuine words as possible.

c Consider now the problem of two substances, such as diamond and graphite, which our evidence suggests are forms of the same chemical – carbon. If this is so, then a crystal of diamond and a flake of graphite must both be built up from the same building block – the atom of carbon. This is a different situation from that in the two examples above, in that we now have one type of 'basic unit', not a number of different types of unit such as the four integers, or the four letters.

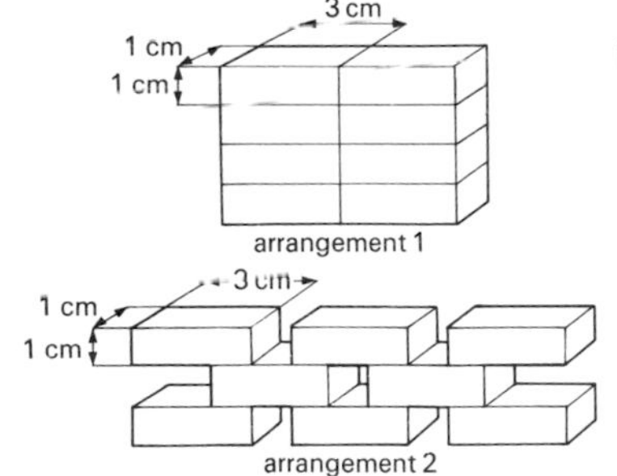

Figure 7.3

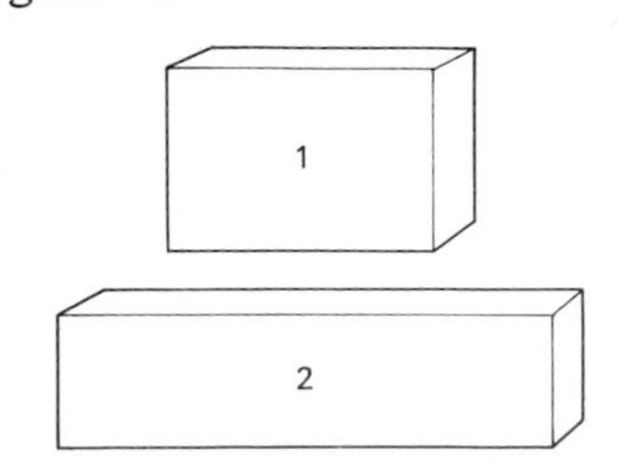

Figure 7.4

However, we can easily see how different arrangements of one type of 'basic unit' can account for different properties. Two arrangements of eight building bricks, each $3 \times 1 \times 1$ cm, are shown (figures 7.3 and 7.4). In arrangement 2 the blocks in each row are separated by a distance of 1 cm.

Imagine that we tightly and smoothly cover both arrangements with paper so that we cannot see the bricks underneath.

►a What is the volume occupied by arrangement 1?◄

►b What is the volume occupied by arrangement 2?◄

►c Explain why arrangement 2 will have a lower density than arrangement 1.◄

In this way a difference in a property, such as density, can be explained in terms of a difference in the arrangement of the building blocks in a substance. Similarly, we might account for the differences in properties of diamond and graphite in terms of differences in the arrangement in space of the atoms of carbon in the structures of these substances.

Crystallographers have examined diamond and graphite using the technique of analysis by X-ray diffraction: the diagrams shown in figures 7.5 and 7.6 represent the structures of the two materials.

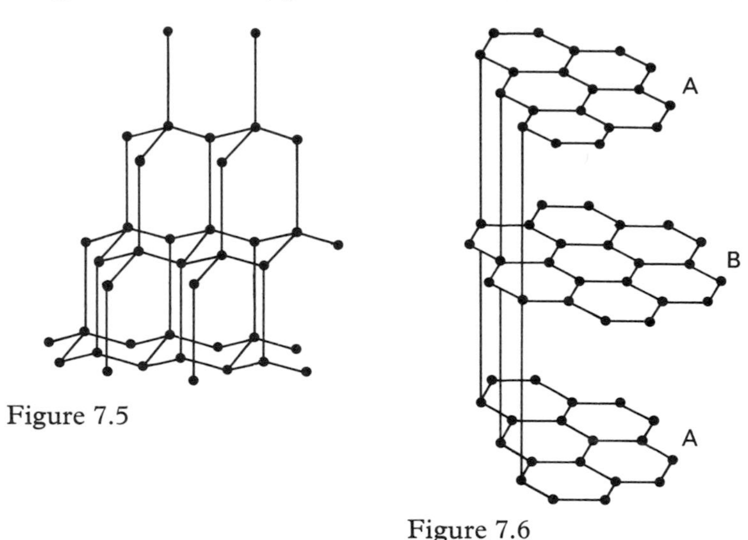

Figure 7.5

Figure 7.6

Construct a model of the diamond structure. This is best done using spheres of expanded polystyrene, but cork spheres and Plasticine spheres can be used quite well. The 'atoms' can be linked together using short lengths of pipe cleaners or cocktail sticks.

The spheres need to have holes pricked in them so that four connectors radiate out from each sphere to the corners of an imaginary regular tetrahedron enclosing the sphere (figure 7.7), so that the angles between the connectors are all equal. This can be done by making a cardboard protractor into which the sphere will neatly fit. If polystyrene or cork spheres 2.50 cm in diameter are being used, a semicircle of radius 1.25 cm is cut out of the card and a mark (↓) made on the protractor to give an angle of 120 °.

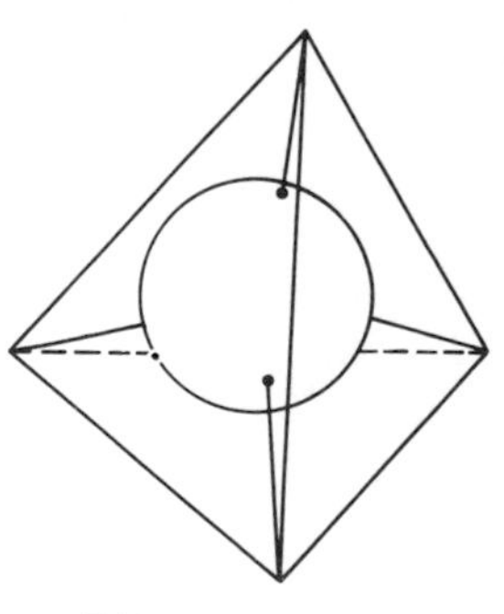

Figure 7.7

A prick is first made at the 'North Pole' (N) of the sphere and

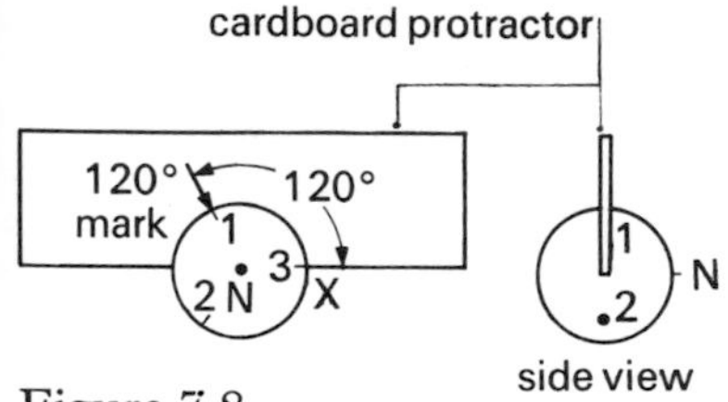

Figure 7.8

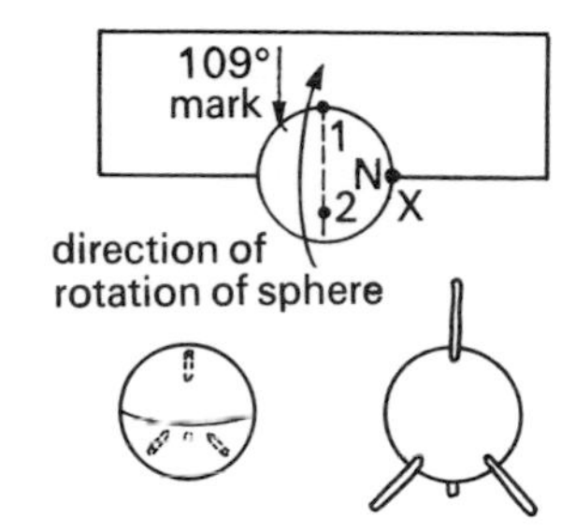

Figure 7.9

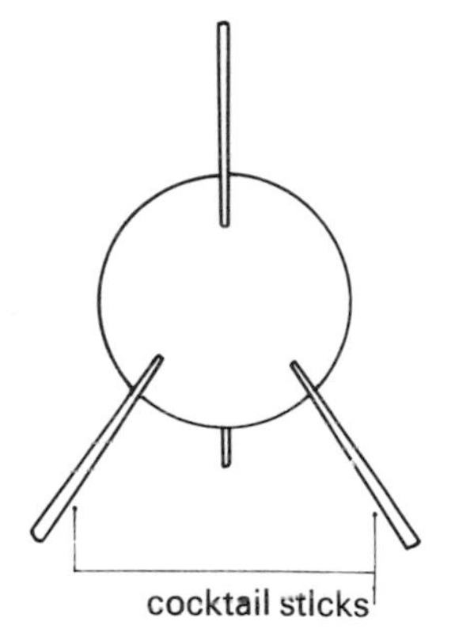

Figure 7.10

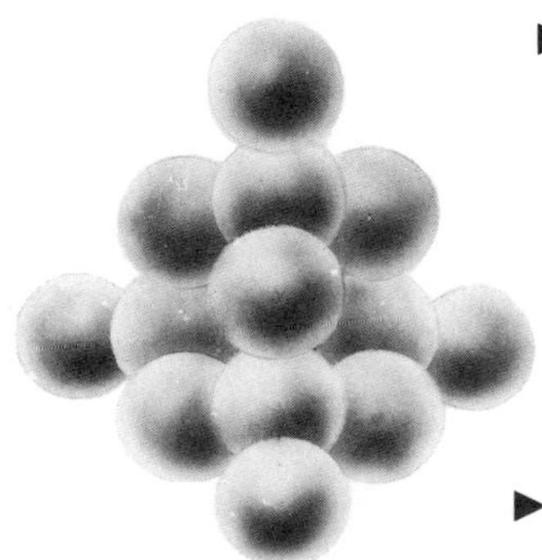

Figure 7.11

three pencil marks are then made on the 'Equator' at 120 ° intervals by turning the sphere whilst in the position shown in figure 7.8. The sphere is then turned through 90 ° so that the 'North Pole' is at point X on the protractor and pencil mark 1 coincides with the curved edge of the semi-circle cut out of the card (figure 7.9). A prick is now made at the 109 ° position. (A compass point or knitting needle pierces the spheres easily.) If the sphere is revolved, keeping N at X, so as to bring pencil mark 2 to coincide with the curved edge of the card, a second prick can be made at the 109 ° position. A third rotation will bring pencil mark 3 into coincidence with the card, and a third prick at the 109 ° position will complete the preliminaries. The sphere is now 'drilled' to take the connecting lengths of pipe cleaner. About 14 spheres should be prepared in this way.

A layer of spheres should next be assembled so that each central sphere is connected to four others. When one layer is completed, you will see that there are two possible ways of building on the next layer. Make sure that the 'holes' in the first layer have a sphere of the second layer over them, otherwise the resulting structure will not be that of diamond.

If polystyrene spheres are not available, attempt to make the model with Plasticine spheres. It will be more difficult to position the links accurately between 'atoms', but quite acceptable models can be made by judging the positions by eye.

►Which of the following descriptions more accurately applies to the structure you have made? Give the reasons for your choice:

a continuous giant structure

b discrete molecule.◄

A model in chemistry is useful and effective only as far as it will explain an observation on the material in question. Let us inquire into the ability of the diamond model to explain some of the properties of diamond:

►i Examine the structure model and consider what happens to the carbon atoms if we attempt to scratch or indent a diamond. Write an explanation of why you think a diamond is very hard – i.e. why it resists deforming and indentation.◄

ii Although diamonds are extremely hard, the skilled craftsmen who cut diamonds to shape make use of the fact that they can be split relatively easily. Look again at the model. ►Explain why there are certain directions and planes along which it might be easier to split a diamond.◄

►iii Into which one of the following shapes would it be most easy to cleave a diamond?◄

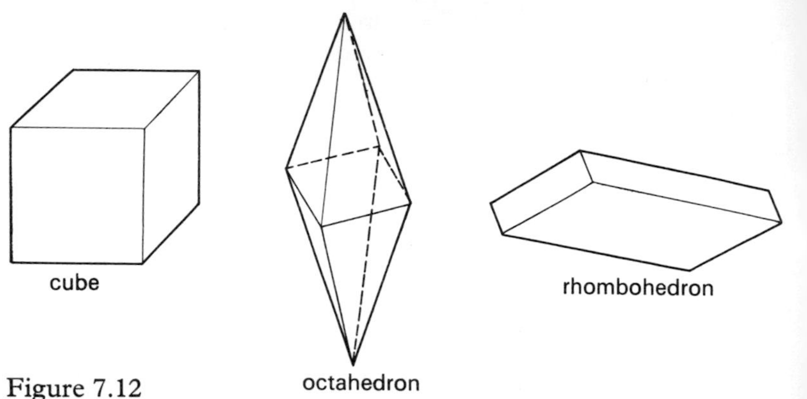

Figure 7.12

iv Diamond cutters know from experience that not all faces of a diamond are equally hard. It is much more difficult to polish certain faces than others. Also the polishing rate is considerably different for different directions of the same face. Scientists still do not fully understand why this should be so. The diagram in figure 7.13 shows the positions of carbon atoms in a single layer of a particular face of a diamond. ►Can you suggest a theory to explain why polishing is much more difficult in the 'hard' direction?◄

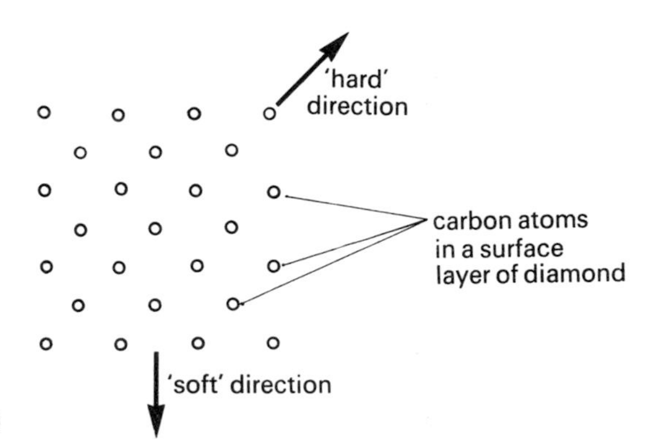

Figure 7.13

v The distance between two bonded carbon atoms in the diamond structure is always the same, i.e. 1.54×10^{-10} m. The distance between bonded carbon atoms in the planes in the graphite structure is 1.41×10^{-10} m but the planes are 3.35×10^{-10} m apart.

►a How does this information help to explain the observed differences in the densities of diamond and graphite?◄ (The density of diamond is 3.5×10^3 kg m^{-3} and the density of graphite is 2.3×10^3 kg m^{-3}.)

b A scientist examines fragments of diamond and graphite, each of which is 0.1 cm^3 in volume. ▶In which of the fragments is there the greater number of carbon atoms?◀

vi In a legend about diamond, reported by Pliny, it is stated that if a diamond is struck a hard blow by an iron hammer, the hammer (and not the diamond) will fly to pieces. Give your views on the credibility of this legend.

▷Investigation 7.6 A study of charcoals

Charcoals are made by heating solid materials such as wood, nutshells and some grades of coal in a restricted amount of air. The starting material is nearly always of biological origin. Although many of the elements in the material are removed on pyrolysis (warming), some hydrogen, oxygen and nitrogen usually remains and charcoals are therefore an impure form of carbon. For example, when wood charcoal is heated to a very high temperature in closed vessels, the resulting material has a higher density than the original wood charcoal and it is a better conductor of electricity and heat than untreated wood charcoal.

Animal charcoal is a term covering the materials which result from the pyrolysis of substances such as bone and blood. Bone charcoal may contain as little as 10% carbon and up to 80% calcium phosphate. Blood charcoal has a much higher percentage of carbon.

Carbon blacks are substances obtained from liquid or gaseous raw materials by methods such as burning in a limited amount of air, playing a flame on a cool surface and the thermal cracking of carbon compounds. Carbon blacks, such as lampblack, usually contain some hydrogen as an impurity, but are reasonably pure forms of carbon.

Gas carbon which is found on the sides of the retorts of gas works, is another form of carbon. It is very hard and almost as pure as lampblack.

Some properties of various forms of carbon are given in the following table:

	wood charcoal	gas carbon	graphite
average density/ 10^3 kg m^{-3}	1.5 (air removed from pores)	1.9	2.3

	wood charcoal	gas carbon	graphite
conduction of electricity	poor	good	good: better than gas carbon
conduction of heat	poor	poor, but better than wood charcoal	good

Because they do not show any crystalline form, it has been suggested that materials such as charcoals, carbon blacks, bone chars, etc., represent a third distinct allotropic form of carbon, termed amorphous carbon. Do you agree with this opinion? Support your answer by reference to the information given above.

The following passage is taken from a scientific paper written in 1844 by two French experimenters, Fizeau and Foucault. The electric arc which they describe was struck between two carbon rods. Provided the gap between the rods is not too large, carbon vapour will allow the current to flow across the gap.

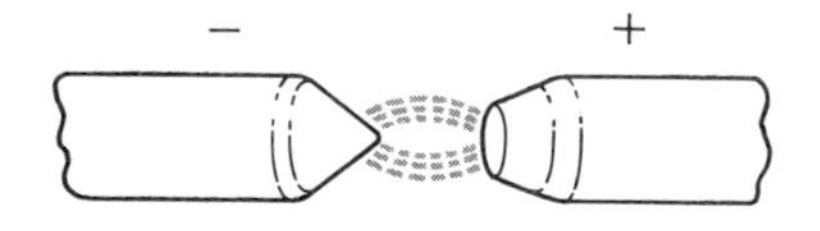

Figure 7.14

Nous terminerons en appelant l'attention sur une modification remarquable éprouvée par le charbon lorsqu'il a supporté la très-haute température qui se développe pendant l'incandescence des pôles de la pile[1].

Le charbon très-dense qui provient[2] de la distillation de la houille[3], et que nous avons employé, a des caractères physiques qui le rapprochent de l'espèce minérale appelée anthracite; en examinant, après les expériences d'incandescence, le charbon transporté au pôle négatif et l'extrémité du pôle positif lui-même, nous avons remarqué que ses caractères physiques sont alors changés.

Ce charbon est mou[4], traçant[5]; sa surface, étant frottée[6], devient d'un gris plombé métallique. Ces caractères l'assimilent[7] complètement à l'espèce minérale appelée graphite; cette modification se fait très-rapidement, et s'obtient également avec d'autres espèces de charbons conducteurs. Il suffit de promener l'arc lumineux sur la surface d'un des pôles de charbon pour que cette surface soit à l'instant revê tue[8] d'une couche[9] de graphite.

Cette formation de graphite, sous l'influence d'une température très-elevée, nous semble devoir jouer un rôle important dans l'étude des masses minérales où se rencontre[10] si fréquemment cette variété de charbon.

[1]la pile *the electric battery*
[2]provenir de *to result from*
[3]la houille *coal*
[4]mou *soft*
[5]traçer *to trace, to mark paper*
[6]frotter *to polish*
[7]assimiler *to liken*
[8]revêtir *to cover*
[9]la couche *coating*
[10]se rencontrer *to be found*

i How would you attempt to obtain, by a simple experimental procedure, an approximate value for the temperature of the arc?

ii To which one of the following properties of carbon is the temperature of the arc related most closely? Give the reason for your choice:

a the melting point of carbon

b the boiling point of carbon

c the electrical resistance of carbon.

iii Fizeau and Foucault write of a very dense form of carbon obtained by the pyrolysis of coal. To which of the varieties of carbon described earlier do you think this refers?

iv Write a simple word equation, including a reference to the conditions of the reaction, to describe the changes which Fizeau and Foucault record.

Investigation 7.7 Crystals of metals

You will need

conical flask, 100 cm^3, or test tube, 150 × 25 mm
test tube, 100 × 16 mm
dilute lead ethanoate solution
strip of zinc foil
thick copper wire, 15 cm long
dilute silver nitrate solution

Suspend the strip of zinc in a solution of lead ethanoate in a flask. Suspend a helix of copper wire in a solution of silver nitrate in a test tube. Describe what happens in each case.

Look for metallic objects at home which are crystalline (e.g. brass handles).

Models of giant structures, made in a similar way to that in Investigation 7.5, can be used in a class discussion to understand the formation of metal crystals.

You now know that atoms can interact to give molecules and giant structures. Elements consist of interacting atoms of a similar type. Compounds consist of interacting atoms of a different type.

Investigation 7.8 Comparing the melting points and boiling points of molecules and giant structures

You will need

a book of data

Use the book of data to find the melting points (m.p.) and boiling points (b.p.) of the elements listed:

element	structure	m.p./K	b.p./K
aluminium	giant		
carbon	giant		
copper	giant		
iodine	molecule		
iron	giant		
phosphorus	molecule		
silicon	giant		
sulphur	molecule		

What pattern does there seem to be? ▶Would you expect elements which are gases at room temperatures to be molecules or giant structures? What about the elements which are liquids at room temperature? Are they exceptions to the pattern? Sodium metal is a giant structure. Is it an exception to the pattern?◀

▷Investigation 7.9 The extraction of sulphur

Sulphur is an essential raw material for many industries. Unfortunately there is a world shortage of the element. Amongst many factors contributing to the heavy demand for sulphur has been a very great increase in world demand for fertilisers, many of which require sulphuric acid for their manufacture.

Much of the sulphur used by industries in this country has been imported from Mexico where it is extracted from underground deposits by a process first developed by Herman Frasch at the end of the last century.

In 1865 in Louisiana, USA, whilst boring for petroleum, prospectors discovered a large sulphur deposit beneath a layer of quicksand about 150 metres thick. This quicksand defeated all attempts to mine the sulphur. Eventually Frasch hit upon the idea of melting the sulphur underground and pumping it to the surface as a liquid.

Superheated water was forced under pressure down a pipe which led to the sulphur deposit: the temperature of the water was about 167 °C. Molten sulphur was then forced up another pipe by means of hot compressed air.

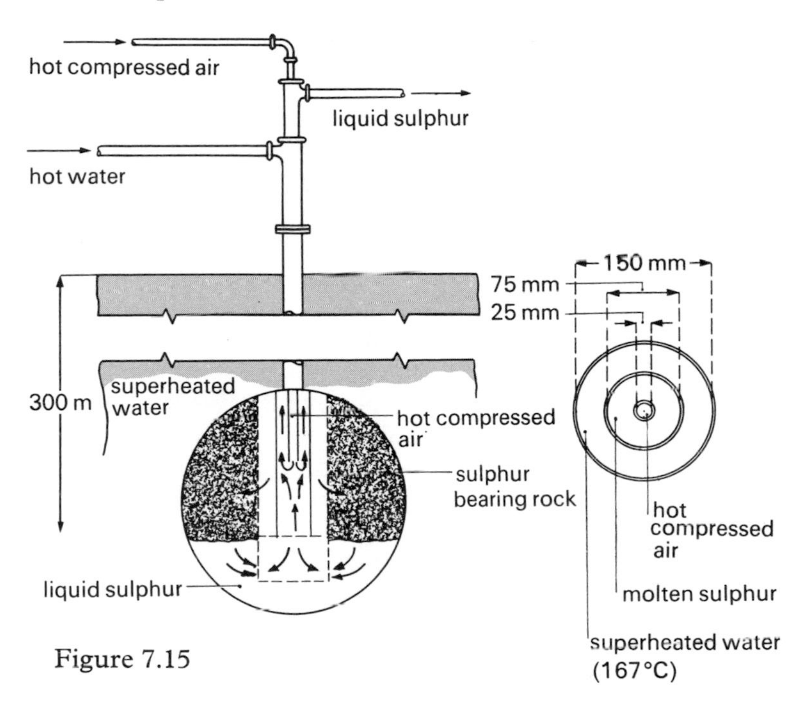

Figure 7.15

Here is Frasch's own account of his pioneer experiment:

When everything was ready to make the first trial, which would demonstrate either success or failure, we raised steam in the boilers, and sent the superheated water into the ground without a hitch. If for one instant the high temperature required should drop below the melting point of sulphur, it would mean failure – consequently intense interest centred in this first attempt.

After permitting the melting fluid to go into the ground for twenty-four hours, I decided that sufficient material must have been melted to produce some sulphur. The pumping engine was started on the sulphur line, and the increasing strain against the engine showed that work was being done. More and more slowly went the engine, more steam was supplied, until the man at the throttle sang out at the top of his voice, 'She's pumping'. A liquid appeared on the polished rod, and when I wiped it off with my finger, I found my finger covered with sulphur. Within five minutes, the receptacles under pressure were opened, and a beautiful stream of the golden fluid shot into the barrels we had ready to receive the product. After pumping for about fifteen minutes, the forty barrels we had supplied were seen to be inadequate. Quickly we threw up embankments and lined them with boards to receive the sulphur that was gushing forth, and since that day no further attempt has been made to provide a vessel or a mould into which to put the sulphur.

When everything had been finished, the sulphur all piled up in one heap, and the men had departed . . . I mounted the sulphur pile and seated myself on the very top. It pleased me to hear the slight noise caused by the contraction of the warm sulphur, which was like a greeting from below . . .

(From *J. Soc. Chem. Ind.* xxxi, 1912, 168).

1 Explain why it was necessary to superheat the water under pressure in order to extract the sulphur.

2 Explain why it was sensible to arrange the pipes so that the molten sulphur was carried upwards in a pipe, the outside surface of which was in contact with superheated water and the inside surface with hot compressed air.

3 In what way did the following factors contribute to the success of Frasch's pioneer experiment for extracting sulphur?

a The relatively low melting point of sulphur.

b The ready availability of an inexpensive fuel (e.g. petroleum) with which to warm the water.

c The mobility of liquid sulphur at temperatures just above its melting point.

4 Frasch writes of wiping a liquid from a rod on the pumping engine. What would be the approximate temperature of this liquid?

5 If you take a stick of roll sulphur from a laboratory bottle, and grip the stick in your warm hand, holding it to your ear, it is possible to hear occasional sharp cracks, similar to the noise heard by Frasch as he sat on his pile of newly won sulphur. Sometimes, without exerting any pressure from the hand, the stick will fracture. Write a brief explanation of the noise and of the fracture.

6 In which allotropic form, rhombic or monoclinic, would you expect the sulphur to crystallise in the barrels? Give your reasons.

7 When molten sulphur is collected in barrels, and cooled to a solid, it is frequently difficult to remove the newly crystallised product without breaking the barrel. The same effect can be seen if an attempt is made to remove freshly-solidified sulphur from an evaporating basin. After some weeks the sulphur can be dislodged from the sides of the container without difficulty. Write a brief explanation of this observation.

Investigation 7.10 Giant structures from different atoms

▶Devise and carry out experiments to discover whether sand (silicon dioxide) and water are molecules or giant structures.◀

In the previous section you calculated the size of a molecule. If molecules are made from atoms, then atoms must be even smaller building blocks!

Investigation 7.11 Calculating numbers of atoms

In Investigation 7.2 you calculated the sizes of uranium and platinum atoms from electron micrographs. Use these figures to calculate the number of atoms placed side by side which would measure 1 cm. From this number work out the number of atoms in 1-cm cubes of uranium and platinum.

What large numbers we are dealing with! The mass of one carbon atom is thought to be about 0.000 000 000 000 000 000 000 02 g, or 2×10^{-23} g! The density of graphite carbon is 2.26 g cm^{-3}. Calculate the number of carbon atoms in a 1-cm cube of graphite.

This investigation will have shown that we are working with very large quantities of extremely tiny particles. Before carrying out further calculations involving such large numbers of atoms you are going to experiment with much larger chunks of stuff: spheres of coloured plastic. These experiments should help you to understand what you are doing when the chunks of stuff are atom-sized particles.

Investigation 7.12 Weighing in 'units'

You will need

ten white spheres
one each of blue, yellow and red spheres
lever balance and scale pans

Use the simple lever balance to find the number of white spheres needed to balance one of the other coloured spheres. Complete the second column of the following table:

colour of sphere	number of white spheres to balance one of the other spheres	mass/'units'
white		
blue		
yellow		
red		

If one white sphere weighs one 'unit' calculate the mass of each of the other spheres in 'units' and complete the third column of the table.

How many times as heavy is a blue sphere as a yellow sphere? How many times as heavy is a yellow sphere as a red sphere?

Investigation 7.13 Weighing bags of spheres

You will need

sealed opaque polythene bags containing equal (but unknown) numbers of:

a red spheres
b blue spheres
c yellow spheres
d white spheres

lever balance and scale pans

In this and all future investigations the polythene bag should be counterbalanced by a second (empty) bag. Use white spheres to balance each of the bags in turn and complete the following table:

colour of spheres	bag mass/'units'	$\frac{\text{mass of bag}}{\text{mass of white bag}}$
red		
blue		
yellow		
white		

Compare the last column with the masses of single spheres in 'units' (Investigation 7.12).

Suggest how the results of this investigation might be useful when dealing with atom particles. The 'standard' you took in this experiment was the lightest sphere (white). What might be a suitable standard for the atom building block?

In this investigation you did not need to know the actual number of spheres in a bag. In the next experiment you will be finding this number.

Investigation 7.14 Finding the number of spheres

You will need

any one of the opaque bags
white spheres
lever balance and scale pans
ruler

Find the total mass (in 'units') of the contents of the bag. From your knowledge of the mass of one sphere (Investigation 7.12) calculate the number of spheres in the bag.

Of course, it is far easier just to count the actual number of spheres, but with atom particles this is not so readily done. (Why not?) However, it can be done, as you will see later! Link what you have done in this investigation with Investigation 7.11.

So far you have worked in 'units'. In the next experiment you will be repeating Investigation 7.13, but you will be weighing in grammes.

Investigation 7.15 Weighing in grammes

You will need

sealed, opaque polythene bags containing equal (but unknown)

numbers of spheres
access to top pan balance

Carry out experiments to complete columns 1 and 2 of the following table:

	1	2	3
bag	mass/g	$\frac{\text{mass of bag}}{\text{mass of bag A}}$	$\frac{\text{mass of single sphere}}{\text{mass of sphere A}}$
A			
B			
C			
D			

To calculate the actual mass of a single sphere in each bag, what further information would you require? If you know the actual masses of single spheres, what would be the ratios in column 3? Remember, each bag contains the same number of spheres.

Investigation 7.16 From spheres to atoms

You will need

transparent polythene bags containing equal numbers of the following (real!) atoms:

a carbon
b calcium
c sulphur
d magnesium

access to top pan balance
book of data

Carry out experiments to complete columns 1 and 2 of the following table:

	1	2	3	4
bag	mass/g	$\frac{\text{mass of bag}}{\text{mass of carbon}}$	mass when C = 12 g	relative atomic mass
carbon				
sulphur				
magnesium				
calcium				

What would be the mass of the other three bags if the mass of the carbon bag were 12 g? Complete Column 3. Would this be true of both 'units' and of grammes? Use your data book to complete Column 4.

What can we say about the number of atoms in bags containing the atomic mass in grammes of an element?

In this experiment you were told that the bags contained an equal number of atom building blocks. It was impossible to count these individually, as with spheres! These masses were calculated from the results of combining one atom with another; you will be making these calculations later in this section. In the next investigation you will be calculating the actual numbers of atoms in the relative atomic mass of an element.

Investigation 7.17 The Avogadro constant

Already you know that one atom of carbon weighs about 2×10^{-23} g. Complete the following table:

element	relative atomic mass A_r	$\frac{\text{mass of 1 atom of element}}{\text{mass of 1 atom of carbon}}$	mass of 1 atom/g	number of atoms in A_r
carbon	12	$\frac{12}{12} = 1$	2×10^{-23}	
sulphur				
magnesium				
calcium				

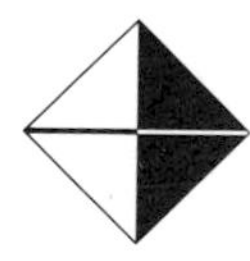

The number obtained in the last column is called the Avogadro constant. The amount of substance containing this number of atoms is given a special name: the mole. What is the mass of 2 moles of magnesium, 1 mole of iron and 0.5 mole of sulphur? What mass of aluminium contains the same number of atoms as 12 g of carbon? The mole can also be used for amounts of any other particles (as well as for atoms).

▷ Investigation 7.18 Counting atoms and measuring volumes

This is a cassette and discussion of Rutherford's alpha particle/helium experiment.

Instead of writing in full '1 mole of copper' or '1 mole of carbon', etc., a shorthand is used. This shorthand is the chemical symbol of the element. In particle interactions Cu represents one mole of copper (not one atom of copper!) and C represents one mole of carbon. It is useful to remember the symbols of the more common elements.

At this stage you might find it helpful to answer some of the investigations about the masses of atoms in the *Patterns* book, *Chemical formulae and equations.*

The idea of the mole will now be used to discover if there is any pattern of combination of copper and oxygen atoms when they interact to give black copper oxide.

Investigation 7.19 Patterns of combination of atoms

Black copper oxide

You will need

hard glass test tube, 125 × 16 mm, with small hole near the closed end
length of rubber tubing with means of connection to the test tube
Bunsen burner and hardboard mat
access to two gas taps
balance
retort stand and clamp
pure dry black copper oxide (about 2 g), analytical grade

gas supply

Figure 7.16

The mass of 1 mole of copper is 64 g and the mass of 1 mole of oxygen is 16 g. What further information do you need to find out how many moles of oxygen combine with one mole of copper?

In this experiment we shall reduce a weighed quantity of pure black copper oxide to copper, and then weigh that. What weighings will you be making? Prepare a table in your book so that you can enter each mass directly you have made the weighings. Now make the first two weighings. Put two or three measures of the pure dry copper oxide in the middle of the weighed tube and weigh it again.

Set up the apparatus as shown in figure 7.16. Carefully turn the gas tap half on: if the gas is turned on too far the copper oxide may be blown out of the tube. After waiting ten seconds (why?), light the gas where it comes out of the tube. Turn the gas down till the flame is only about 2 cm high. Warm the tube gently with a small Bunsen flame.

►Two pieces of evidence will tell you that an interaction is occurring. What are they?◄ When the interaction seems to have finished, stop warming and let the copper cool with the stream of gas still passing. Why is this necessary? Turn off the stream of gas and when the whole tube is cool, weigh it again.

What mass of copper was left? What mass of oxygen has gone? ►Write a clear statement saying: '. . . grammes of oxygen were combined with . . . grammes of copper.' How many grammes of copper were combined with 1 gramme of oxygen? (Remember that 1 mole of oxygen is 16 grammes.) How many moles of copper is this? (Remember that 1 mole of copper is about 64 grammes.) How can you represent black copper oxide using symbols? Is black copper oxide a giant structure? What does the formula of black copper oxide actually tell you?◄

Similar patterns of combination can be obtained for other elements. Examples are given in the next investigation.

Investigation 7.20 Formulae of compounds

This is printed as question 4.5 in *Chemical formulae and equations.*

Do you think that the chemical symbols chosen for the elements are sensible?

So far the atoms, molecules and giant structures which have been considered have mainly been from non-living systems. The building blocks of cells are atoms and molecules. Since there are many different atoms and molecules it is only possible to search for the presence of a few of them in organisms. Therefore, pattern-searching must be restricted to just two examples. Nitrogen is a convenient atom and proteins are convenient molecules.

Investigation 7.21 Testing for nitrogen

You will need

ignition tubes
test tube holder
Bunsen burner and hardboard mat
spatula
indicator paper
access to a fume cupboard
samples of common plant or animal materials like meat, fish, beans,

grass, horn and hair
sodalime

Ammonia molecules are made up from nitrogen and hydrogen atoms. A mole of ammonia is represented by NH_3. If a substance containing nitrogen is warmed with sodalime, ammonia molecules are made. This is found to work with almost any substance containing nitrogen. Ammonia is easy to detect, even in small quantities. Hold some damp litmus paper over a bottle containing a dilute ammonia solution. Note what happens.

Mix some ground-up plant or animal material with twice its volume of sodalime in an ignition tube. Each member of the class could test something different. Warm strongly and test with damp litmus paper held near the mouth of the tube. Is ammonia (and therefore nitrogen atoms) present? As the fumes have a particularly unpleasant smell, the experiment should be performed in a fume cupboard or near an open window. Discuss the results of the entire class. Does there seem to be any pattern?

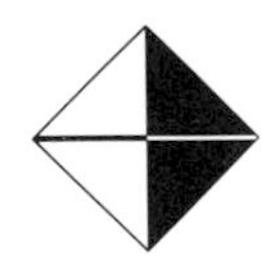

You have probably heard of proteins in connection with food. It would be reasonable to assume that protein molecules are necessary in our diet because they are building blocks of organisms. This is an assumption we can easily test.

Investigation 7.22 Testing for protein

You will need

test tube, 100 × 16 mm
pestle and mortar
Bunsen burner and hardboard mat
albumen (or casein powder)
Millon's reagent
plant and animal materials

Albumen and casein are proteins. Gently warm a little albumen powder with 2 cm^3 of Millon's reagent in a test tube and note the colour change. (Note. Use Millon's reagent with care and wash your hands after using it. It is poisonous.)

Grind up small pieces of material from organisms in a little distilled water. Those used in the previous investigation could be used. Test with Millon's reagent as above. Is the assumption (which is really a pattern) justified?

Consider the two patterns which you have discovered in the last two investigations and the materials used in both sets of experiments. ▶Suggest a possible relationship between nitrogen atoms and protein molecules. Test your suggestion. Discuss ways in which you might extend your search for patterns in organisms, molecules and atoms if you had unlimited time.◀

Investigation 7.23 Atoms, molecules and giant structures: for good or ill

In this section you have learnt that atoms can interact to give molecules or giant structures. You have seen that there are patterns in the interactions and that atoms and molecules are building blocks of cells. Scientists are able to make many thousands of different molecules and giant structures. Most of these are of benefit to man: others can be injurious.

Examine and discuss the photographs (figure 7.17a–c) which show a few of the uses and misuses of chemicals.

Figure 7.17a
Using gas in World War 1

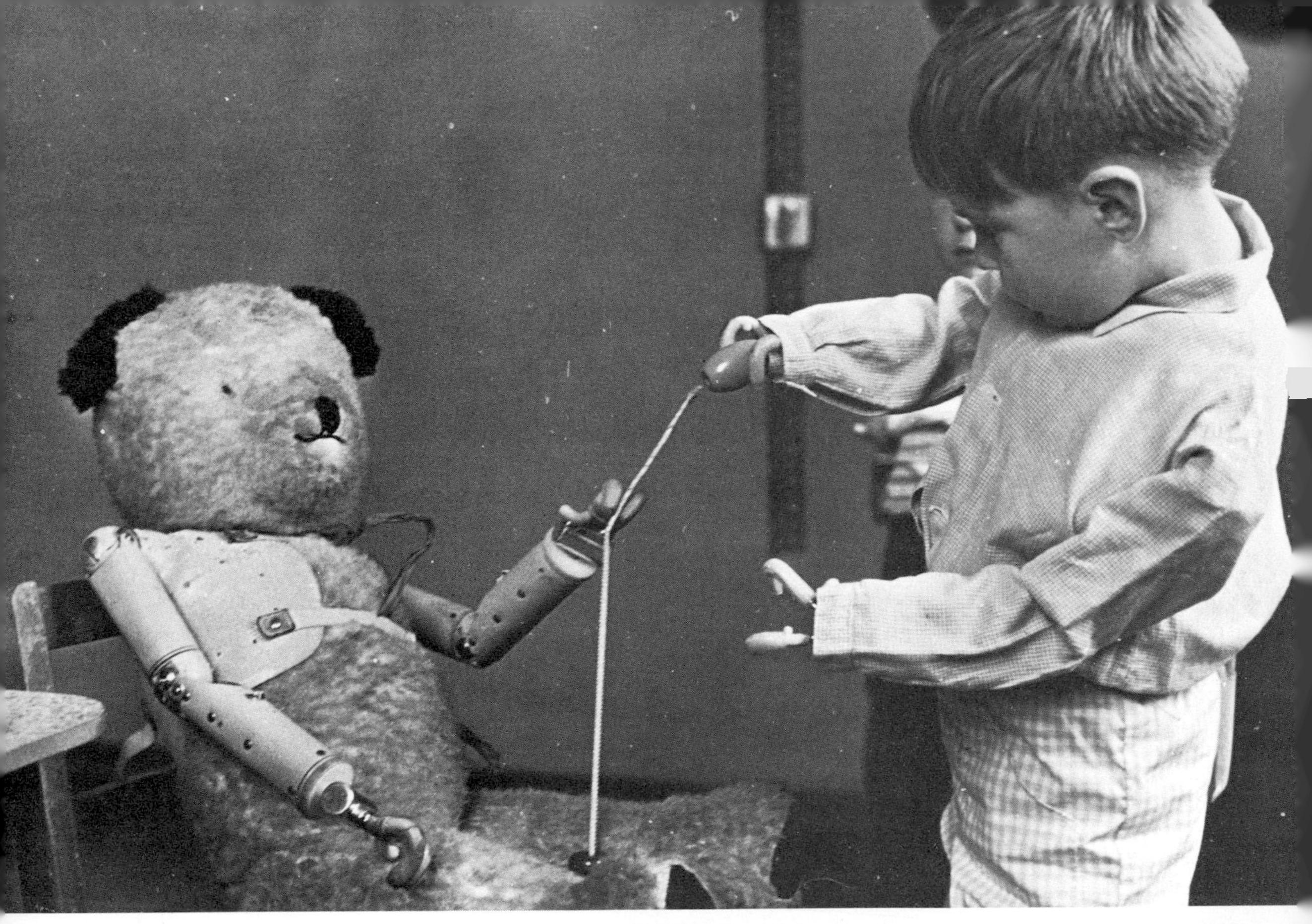

Figure 7.17b
A thalidomide victim

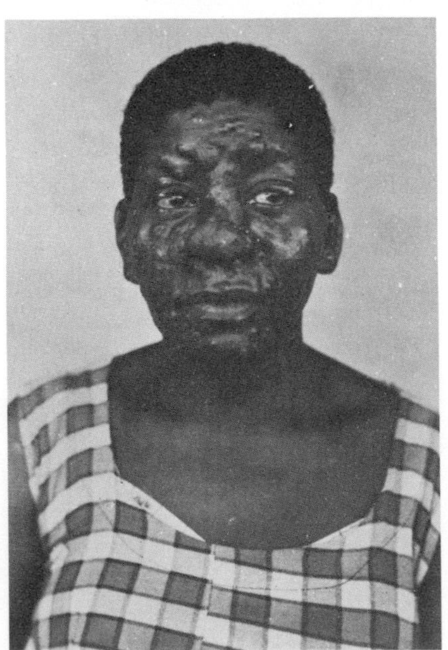

Figure 7.17c
Curing leprosy

8 The electron, ions and giant structures

When nylon clothing is being removed have you heard a crackling sound? Have you ever picked up pieces of paper with a comb after combing your hair? And has your hair ever stood on end after it has been combed? After leaving a plastic chair have your fingers felt an electric shock? Rub your pen or comb on your sleeve and then draw it past the top of your ear, not quite touching it. You should be able to feel the little sparks as the plastic discharges to your ear. In all of these cases there have been forces of attraction and repulsion. In this section you will be taking a closer look at the building blocks which are involved in this.

Investigation 8.1 Electrostatic forces

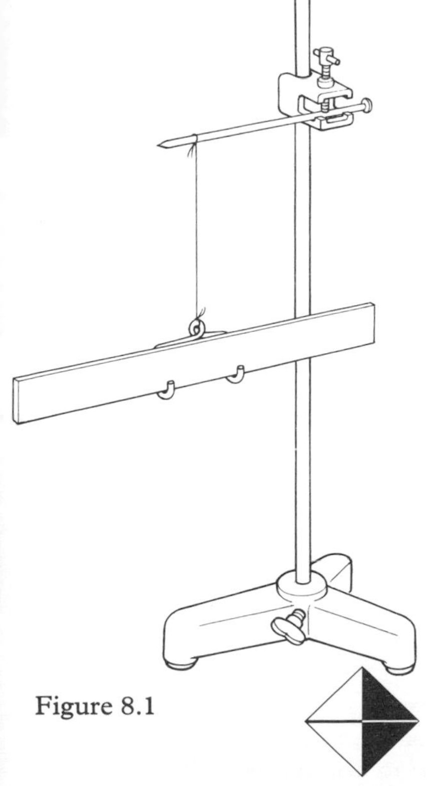

Figure 8.1

You will need

two cellulose acetate strips
two polythene strips or rods
suspending stirrup on nylon thread
retort stand
woollen cloth
Bunsen burner

Hold the polythene strip in the middle (between a thumb and index finger) and rub the whole length of it with the wool. Hang the strip up in the stirrup (see figure 8.1), taking care not to touch the ends of the polythene. Rub another polythene strip with the wool and bring it near to the suspended one. Try all parts of the strips. Now discharge the strips and cloth by passing them over the top of a small yellow Bunsen burner flame. Can you suggest why this is important? It should always be done before starting a new part of the investigation.

Repeat the experiment, but this time rub only the strip you hang up. Bring an unrubbed polythene strip near to it. Is the effect the same as with two rubbed polythene strips? What happens if instead of the unrubbed strip you bring your hand near to the suspended polythene? Try the same thing with an unrubbed acetate strip near to the suspended polythene. What is the pattern of behaviour between unrubbed substances and rubbed polythene?

Repeat all the previous experiments using acetate strips instead of polythene.

Repeat the experiment using one strip of polythene and one of acetate, both of which have been rubbed. Each strip in turn can be suspended. It may be a good idea to discharge the cloth after each rubbing. A pattern of attractions and repulsions should now be emerging: make a note of what you have found.

Rub a polythene strip with wool, suspend it, but this time put the wool near the suspended strip. Do *not* discharge the wool first. Repeat using an acetate strip. This should enable you to extend the pattern and say something about what happens when the wool is used to rub the strips. Discussion with your teacher will extend the ideas of charge, charging and discharge.

So far you have deliberately used only three substances, polythene, cellulose acetate and wool. If other substances are used there could be new possibilities for attraction and repulsion. For example, is there a substance which when rubbed with a suitable material will repel both charged polythene and charged acetate, or attract both? The next investigation will help in answering this.

Investigation 8.2 How many sorts of charge?

You will need

apparatus used in the previous investigation
rods or strips of various other materials (e.g. glass, polystyrene, PVC, wood, ebonite, copper, brass)
various other cloths (cotton, silk, nylon, etc.)

Not all of the objects will become charged when rubbed with cloths. What would you expect to happen with one charged and one uncharged object?

Use all the materials available to answer, as far as possible, the question 'how many sorts of charge?' Is it correct to say that if an object becomes charged it always behaves either like charged polythene or like charged acetate? Are there only two ways in which a substance can be charged? Although you have used only a selection of substances, you should have a pattern. What is the pattern?

We say that, when rubbed with wool, polythene becomes charged negatively and that acetate becomes charged positively. These terms are used because they help to remind us of the pattern of attractions and repulsions. But in the eighteenth century the terms 'resinous' and 'vitreous' were used for what we now call negative and positive respectively. Resin easily acquires a negative charge and glass a positive one when rubbed. The terms negative and positive were first used in this context by Benjamin Franklin in the mid-eighteenth century.

Investigation 8.3 Making a prediction

▶If a woollen cloth is rubbed on polythene what do you predict would happen when the cloth is brought near a charged acetate strip? Test your predictions.◀

All the results so far can be explained in terms of something we call 'charge' being transferred from one substance to another when the substances are rubbed together.

When substance A becomes positively charged, how many possible ways are there for charges to move between A and B to bring about this result? Draw diagrams to illustrate your answer.

Figure 8.2
This machine is used for coating photographic film. After the coating has dried the film is wound on to reels, but the friction tends to charge the film and could cause sparks. The static eliminator (the tube with large holes through which points project) discharges the film and prevents any sparks

Bodies can be charged more easily than by rubbing. The van de Graaff machine (see figure 8.3) is fairly complicated, but you can think of it as a device for carrying charges from the base to the dome.

▷ Investigation 8.4 The van de Graaff machine

This is a demonstration experiment. You can estimate the potential difference (or voltage) produced by measuring the spark, calculating on the basis that each millimetre of spark requires approximately 3 000 volts. This calculation also applies to the sparks from your pen or your shirt.

It seems as if at least one sort of charge must be 'loose' so that it can be transferred easily from one substance to another. The idea of charges moving has probably been mentioned to you before in the context of electric current. It is interesting to see whether there is any connection between the two.

Figure 8.3
A van de Graaff generator

Investigation 8.5 Moving charge and current

This is a demonstration experiment.

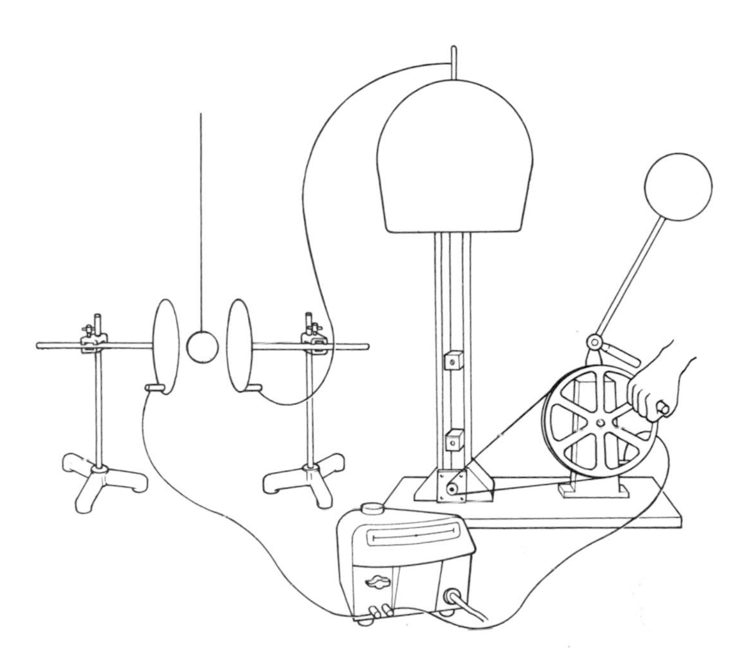

Figure 8.4

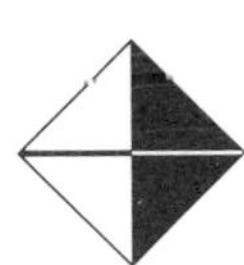

Moving charges give an explanation of an electric current. The charged objects which move are called charge carriers. A continuous transfer of charge is an electric current.

So whenever you meet an electric current (that is, whenever something conducts electricity) you are entitled to ask questions about what charges are moving. Sometimes these questions may not be very easy to answer.

The charge in Investigation 8.5 is transferred in distinct 'chunks', one each time the ball hits the second plate. This raises the question of whether charge exists in 'chunks' of a certain size, or whether it exists as a continuous 'fluid'. Benjamin Franklin thought in terms of an electric fluid and imagined glass as acquiring an extra amount of this fluid when it was positively charged. We now think in terms of electrons, but the evidence for this requires a knowledge of work not yet covered, so we must postpone it until later. This is perhaps not surprising when we remember that the original experiments showing the size of the basic unit of electric charge were done in the early part of the twentieth century. Meanwhile we will accept the name electron for this building block and take it for granted that it is negative.

In the next investigations you will be looking at the transfer of charge through various media.

Investigation 8.6 Charge carriers in a flame

This is a demonstration experiment. ▶Can you suggest why the two parts of the divided flame are different?◀

▶Having seen that a flame contains both negatively and positively charged particles, can you explain how both charged polythene and charged acetate strips can be discharged by the same process, namely passing them over a flame?◀

Investigation 8.7 Charge carriers in solids and solutions

Part a

You will need

d.c. ammeter, 0 – 1 A
U2 cell
circuit board and connectors
plastic covered copper wire
crocodile clips
pea bulb and holder
push switch
test probes (e.g. graphite rods)
beaker, 100 cm^3
zinc
zinc sulphate
lead
lead bromide
starch
sugar
copper
copper (II) chloride
sodium chloride
distilled water

Figure 8.5 shows how to arrange the apparatus. ▶You should be able to predict whether there is a transfer of charge for three of the substances. Check your predictions.◀ Is there a transfer of

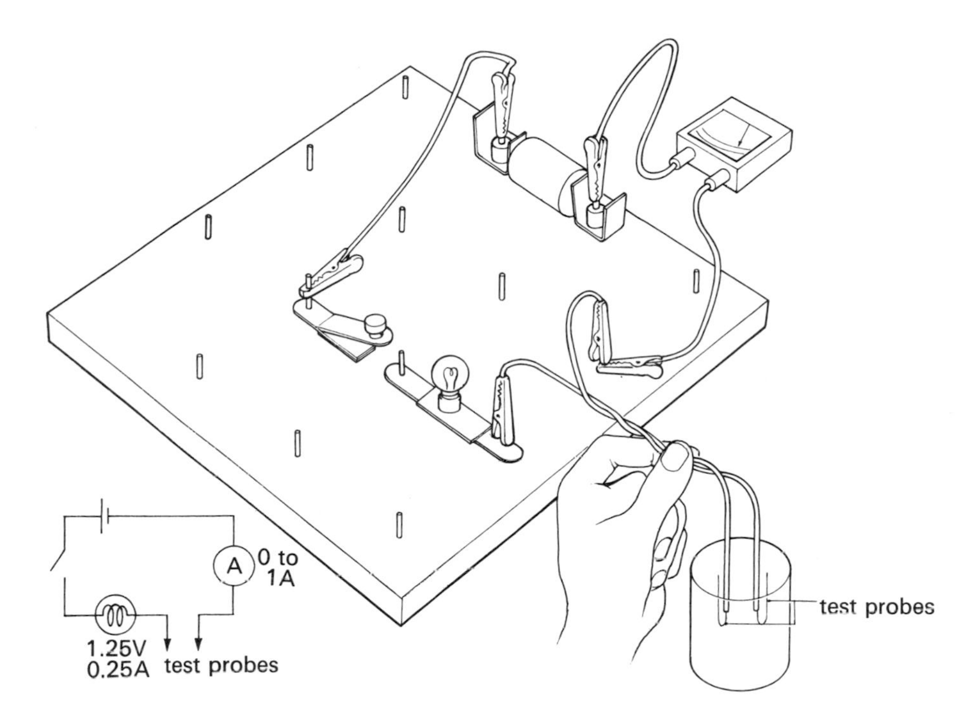

Figure 8.5

charge for the other materials (including distilled water)? Try to dissolve samples of each of the materials in turn in about 50 cm^3 of distilled water. Is there charge transfer through these solutions?

You can now classify the substances into three groups. Summarise this pattern.

When there was charge transfer did other things also happen at the electrodes? You will be considering this in Part c.

Part b

You will need

paraffin wax
Perspex chips
foamed polystyrene
suitable solvents
distilled water

Does paraffin wax fit into the classification? Use a suitable solvent to dissolve the wax and test the effect of electricity on the solution and also on the solvent chosen. Now try the other solids. ▶Does the pattern from part a need modifying?◀

Part c

Take a closer look at the effects of electricity on the conducting solutions in Part a. Identify any gases evolved. What do you observe happening

a at the cathode

b at the anode?

Is there a pattern here? ▶ Can you offer an explanation for this? ◀ Perhaps we should take an even closer look at water?

Investigation 8.8 A closer look at water and salt solution

This is a demonstration experiment. If more sensitive apparatus is used is it possible to detect electricity being conducted by distilled water? What happens when a few grains of salt are added? What happens when the solution is stirred?

For electricity to be conducted between the electrodes in Investigation 8.7 there must have been charge carriers in the solutions. The next investigations provide evidence for the nature of these carriers.

Investigation 8.9 Conduction by coloured salts

You will need

microslide
filter paper strips cut to fit the slide
two lengths of connecting wire fitted with crocodile clips
source of about 24-volt d.c.
small crystal of potassium manganate

Make a filter paper just damp with tap water and place one small crystal of potassium manganate in the centre of it. Apply the voltage between the ends of the paper strip and leave it on for at least ten minutes. Is any movement of the coloured boundary visible?

A similar experiment will be performed as a demonstration for copper chromate.

Examine bottles of potassium and copper compounds as well as bottles of manganates and chromates. Is there any pattern in the colours you see?

▶What explanation can you offer for the observations in this experiment?◀ How can we explain the behaviour of salts? Discussion with your teacher will provide an explanation which is summarised below:

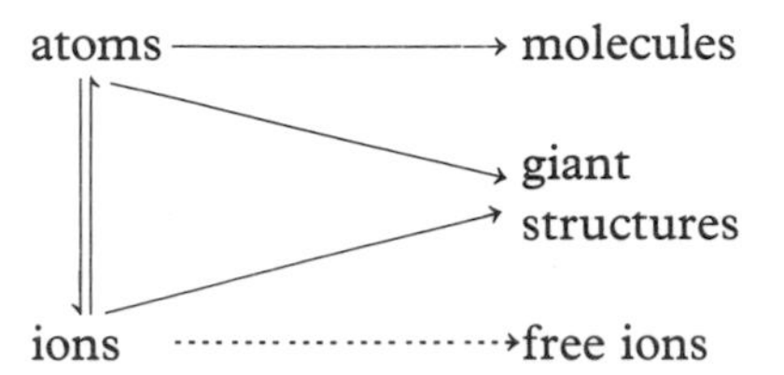

What charges do the ions carry? Can you suggest how the ions are formed?

Investigation 8.10 Charge carriers in molten salts

▶What would you expect to happen when molten lead bromide is electrolysed? Devise an experiment to test your prediction.◀

▶What would you expect to happen when molten sugar is electrolysed?◀ (Molten sugar is not a salt!)

Investigation 8.11 A more detailed look at what happens at the electrodes

You will need

6 volt d.c. supply
beaker, 100 cm^3
rheostat
ammeter, 0.5 A maximum
connecting wire
access to a balance
two copper foil electrodes and support
copper sulphate solution, approximately 0.5 M
distilled water
ethanol
propanone

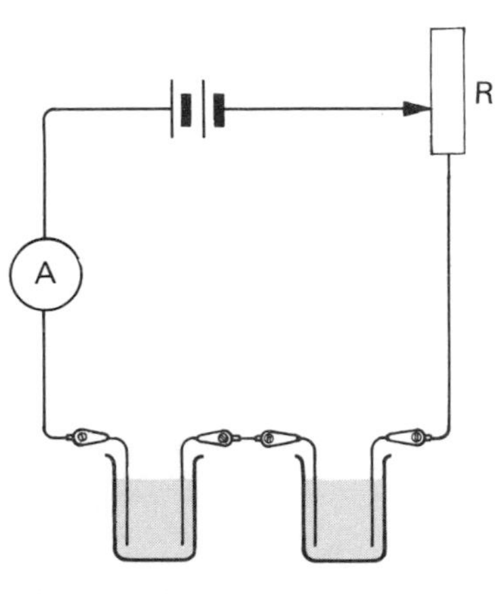

Figure 8.6

Ensure that the electrodes are clean and dry. Weigh them and record their masses. Connect up the electrical circuit as shown. Adjust the rheostat to give a reading of about 100 mA. After one hour, remove the electrodes and wash them by dipping them in distilled water, then in ethanol and finally in propanone. When they are dry, weigh them.

Have the electrodes increased or decreased in mass? How do the two differences in mass of the electrodes compare? How can this be explained?

Investigation 8.12 An application of electrolysis

Read the extract reprinted by courtesy of *Engineering* :

> The inside of the tanker (figure 8.7) was polished by Electropol Processing Limited. The process took about six hours using an electric current of 1 000 A at 10 volt. The electrolyte was a mixture of sulphuric acid and phosphoric acid.

Figure 8.7

Electro-Polishing of Stainless Steels

Four companies have related their experiences with electro-polishing of stainless steel, confirming some of the technical advantages claimed for the process compared with mechanical polishing.

MANY a decision to use stainless steel for a fabrication would be made easier if the polishing operation were more straightforward. Mechanical polishing is sometimes just not practicable. What is more, the high cost of polishing has to be added to the already high cost of the basic alloys. It is understandable, therefore, why bright-anodized aluminium is presenting stainless steel with such strong competition in such fields as architecture.

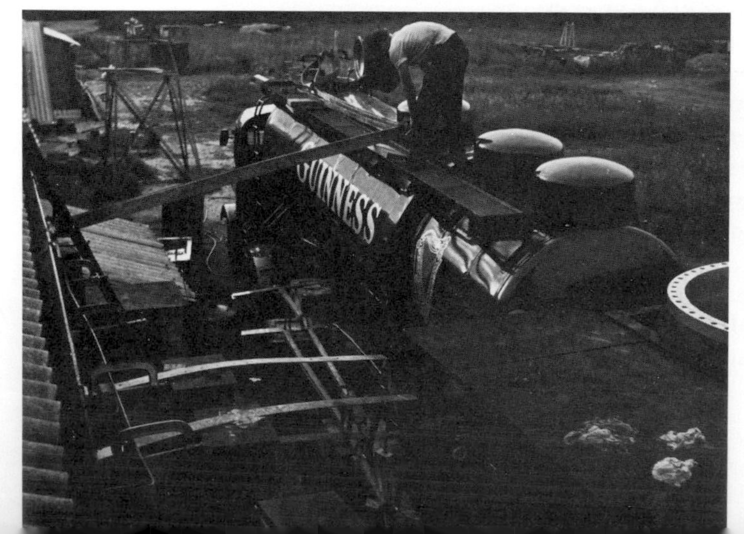

Mechanical polishing is not the only method of finishing stainless steel. The alternative—electro-polishing—has been known and practised for the past fifteen years, certainly by the firm who claim to be the first and still the largest firm of electro-polishers in the United Kingdom, Electropol Processing Limited, Farnham. This firm operates under a sole external jobbing licence from the patent-holding company, Electropol Limited.

Besides claiming that electro-polishing enables whole fabrications to be polished which could only be done by hand before assembly, Electropol Processing also claim that it is a cheaper process.

The Electropol process does not normally produce a mirror finish, but in removing about one-thousandth of an inch from the surface of an article, it levels out any fine scratches. Something very near a mirror-finish has been produced in early trials on bright-annealed stainless steel sheets.

Stainless steel items up to 20 ft long can be polished by Electropol Processing, and the company say they will shortly have a bath 4 ft longer which should be particularly suitable for architectural sections.

There does not appear to be any maximum size which can be handled other than those set by the tanks and lifting apparatus. The company handle electrolyte by the thousands of gallons. According to them, the electrolyte has such good throwing power that whole fabrications can be polished without any costly electrode design. It should also be possible to undertake electro-polishing on site where the practical difficulties of transporting the electrolyte and pumping it in to and out of vessels (often high above ground level) can be surmounted. Most work is still performed at the Farnham works of the company.

►In the case of the tanker explain:

a where the electrolyte would be poured
b where the cathode would be
c where the anode would be
d how the inside became polished.◄

What objects do you have in the home which have been electroplated?

Are ionic solutions charged or neutral?

Atoms (or groups of atoms) can form ions. Ions are charged positively or negatively. It is ions which are the charge carriers both in aqueous solution and when salts are molten. In both cases the ionic solid giant structure is broken down, enabling the ions to move around. The next investigation enables us to learn more about the nature of ions.

Investigation 8.13 Size of ions

The following table shows the sizes of atoms of thirty-three elements (complete atoms are, of course, uncharged), and alongside each the size of the ion formed from that atom, and the sign of its charge.

atom	uncharged radius/10^{-10}m	ionic radius/10^{-10}m	nature of charge (+ or −)
aluminium	1.20	0.51	+
antimony	1.40	2.45	−
argon	1.74	–	none
arsenic	1.19	2.22	−
beryllium	0.90	0.35	+
boron	0.84	0.23	+
bromine	1.14	1.96	−
calcium	1.39	0.99	+
carbon	0.77	2.60 0.16	see problem below
chlorine	0.99	1.81	−
fluorine	0.71	1.33	−
gallium	1.26	0.62	+
germanium	1.22	0.53	+
helium	0.93	–	none
hydrogen	0.37	1.54	−
indium	1.44	0.81	+
iodine	1.33	2.20	−
krypton	1.89	–	none

atom	unchanged radius/10^{-10} m	ionic radius/10^{-10} m	nature of charge (+ or −)
lithium	1.34	0.68	+
magnesium	1.22	0.66	+
neon	1.31	–	none
nitrogen	0.75	1.71	−
oxygen	0.73	1.32	−
phosphorus	1.11	2.12	−
potassium	1.96	1.33	+
rubidium	2.06	1.47	+
selenium	1.16	1.91	−
silicon	1.17	0.42	+
sodium	1.54	0.97	+
strontium	1.49	1.12	+
sulphur	0.03	1.84	−
tellurium	1.35	2.11	−
tin	1.41	0.93	+

Compare the uncharged radius with the corresponding ionic radius. Is there any pattern here? Some atoms do not form ions easily (see table). Do they form a pattern?

▶The last column is incomplete in the case of carbon. Predict the nature of the charge for both sorts of carbon ion.◀

▶In addition to the ion of radius 0.93×10^{-10} m shown in the table, tin forms another sort of ion of radius 0.71×10^{-10} m. What can you suggest about this ion?◀

Both atom and ion building blocks can form giant structures. Atom building block giant structures were investigated in Section 7. You will now look at ion giant structures.

▷ Investigation 8.14 Examining crystals

You will need

hand lens
access to a microscope
variety of crystals

Use the hand lens or microscope to examine the crystal specimens. Do any of the shapes occur more than once? Draw the different shapes.

▷ Investigation 8.15 Looking for crystals at home

You will need

large selection of solid household materials
hand lens

Examine a wide range of solid household substances and list those which are crystalline. How big are the crystals? Do they dissolve in water? What happens if they are exposed to the atmosphere for 24 hours?

Parts a and b of the next investigation illustrate the pattern that acids and bases interact to give salts. In all three parts of the investigation the acid is neutralised.

Investigation 8.16 Making salts

You need do only one of these experiments. Experimental observations and results can then be discussed with those members of the class who have performed the other investigations.

▷ Part a Copper sulphate

You will need

beaker, 100 cm^3
Bunsen burner, tripod and gauze
evaporating basin
filter funnel and filter paper
dilute sulphuric acid 1 M
black copper oxide

Warm 50 cm^3 of dilute sulphuric acid almost to boiling point in a beaker. Add one measure of copper oxide to the acid and boil gently for about a minute. If none of the copper oxide remains, add more until you have excess.

Filter off the excess copper oxide and collect the filtrate in an evaporating basin. Warm the basin and contents until the copper sulphate solution is reduced to about half of its original volume.

Allow the solution to stand overnight. Pour off the excess solution from the crystals and dry them between filter papers. Describe the colour and shape of the crystals.

▷ Part b Sodium chloride

You will need

measuring cylinder
beaker, 100 cm^3
stirring rod
Bunsen burner, tripod and gauze
evaporating basin
filter funnel and filter paper
dilute sodium hydroxide solution
dilute hydrochloric acid
litmus solution
small pieces of charcoal

Measure 25 cm^3 of dilute sodium hydroxide solution (take care!) and pour into the beaker. Add three drops of litmus solution. Slowly add dilute hydrochloric acid solution to the beaker, continually stirring with the glass rod, until the litmus changes to purple. (If too much acid is accidently added, slowly pour sodium hydroxide solution into the beaker until the litmus turns purple.) Boil the neutral solution with some small pieces of charcoal until the solution is colourless. Filter off the charcoal and collect the filtrate in an evaporating basin.

Warm the evaporating basin until most (but not all) of the water has boiled away. Allow the basin and contents to cool. Can you identify any crystals? What colour are they?

▷ Part c Zinc sulphate

You will need

beaker, 100 cm^3
Bunsen burner, tripod and gauze
evaporating basin
filter funnel and filter paper
dilute sulphuric acid
granulated zinc

Pour about 25 cm^3 of dilute sulphuric acid into the beaker. Add five or six large pieces of zinc. Warm the solution gently. Identify the gas which is given off.

When the reaction has stopped, filter off the excess zinc and collect the filtrate in an evaporating basin. Boil the solution until half of the water has been removed. Allow the solution to stand overnight. Pour off the excess solution, examine the crystals and describe their shape and colour.

▷Investigation 8.17 Watching crystals grow

This is a demonstration experiment where you can watch crystals being formed from solutions.

Figure 8.8
Some natural crystals

Investigation 8.18 Using polystyrene spheres to understand crystal shape

You will need

foamed polystyrene spheres

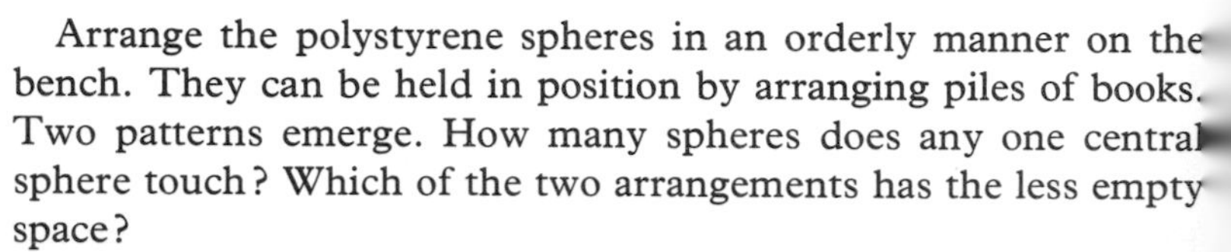

Arrange the polystyrene spheres in an orderly manner on the bench. They can be held in position by arranging piles of books. Two patterns emerge. How many spheres does any one central sphere touch? Which of the two arrangements has the less empty space?

Now try to make three-dimensional models. There are four ways of packing which you should discover, two of which are illustrated in figure 8.9.

▶Are any of the shapes similar to the crystal shapes which you have seen? If you assume that crystals are also made up of ions, can you explain the crystal shapes which you have seen?◀

▶How are ion building blocks held together in an ionic crystal? How are ion building blocks kept apart in an ionic crystal? Are the properties of an ion similar to the properties of the uncharged atom?◀

▶Is it possible to have a bottle of (say) sodium ions separate from a bottle of chloride ions?◀

You have already discovered that one of the elements present in protein is nitrogen. If there had been unlimited time in your previous work you would have been able to discover the presence of other elements such as sulphur in both plant and animal protein molecules. In fact organisms contain a wide variety of elements in the molecules from which they are constructed (see following table).

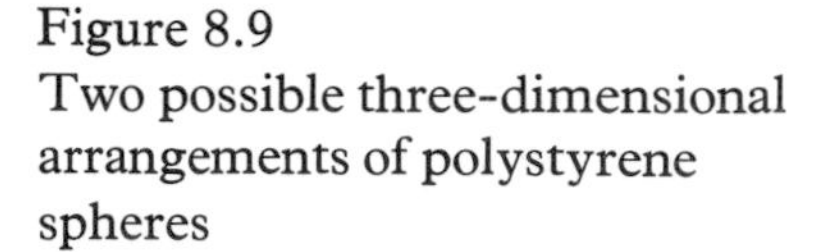

Figure 8.9
Two possible three-dimensional arrangements of polystyrene spheres

Analysis of plant tissues for certain elements, expressed as grammes per 100 grammes dry mass of shoot tissue.

	K	P	Ca	Mg	S
onion	2.6	0.19	2.6	0.3	0.18
barley	3.9	0.35	0.7	0.2	0.19
chickweed	3.7	0.4	1.32	1.7	0.14
mustard	3.1	0.4	1.8	0.3	0.6
broad bean	2.1	0.27	0.9	0.2	0.14
dandelion	4.4	0.49	1.0	0.3	0.22
Elodea	2.2	0.5	5.5	0.8	0.55
water plantain	1.8	0.24	1.0	0.3	0.39

When studying the atom and molecule building blocks of organisms you discussed the idea that we eat proteins so that they can be built into our own bodies – they are amongst the molecule building blocks from which we are made. Look back at your notes on the materials you tested for the presence of protein molecules. Some came from animals and some from plants. Like ourselves, the animals from which the materials came could obtain their protein molecules from plants or other animals. But how do plants obtain their protein building blocks, which we know they contain?

There are two possibilities:

a they can obtain them as proteins, direct from their surroundings

b they can make them from smaller building blocks such as other molecules, or ions, obtained from the surroundings.

How can possibility (a) be tested?

To test possibility (b) you must look for evidence that plants can make proteins from simpler building blocks.

Investigation 8.19 Plant growth

ivy-leaved duckweed

gibbous duckweed

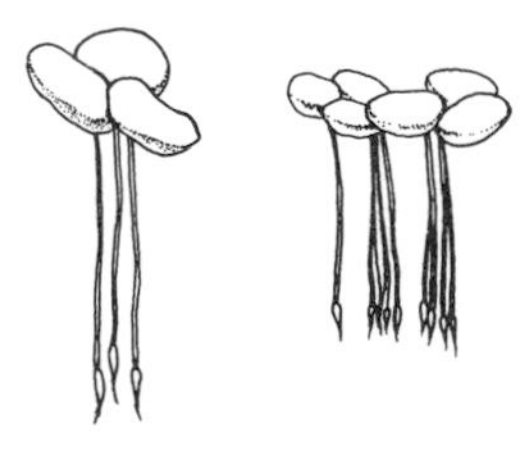

lesser duckweed

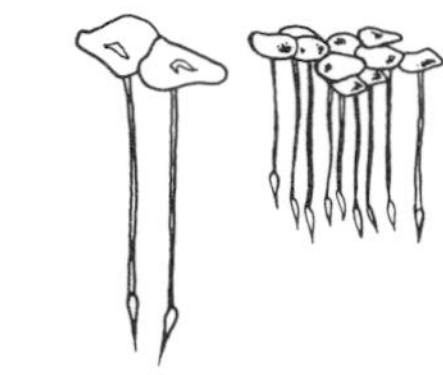

Figure 8.10
Lemna sp.

You will need

dish A and dish B with growing *Lemna* plants
sample of pond water known to grow *Lemna*
Bunsen burner, tripod and gauze
watch glass
beaker
dilute hydrochloric acid
dilute nitric acid
silver nitrate solution
barium chloride solution

Since proteins appear to be important molecule building blocks of organisms you might justifiably assume that if growth takes place then proteins are being obtained or made.

The aquaria in which you are searching for patterns of change in communities may contain small green plants called *Lemna* (duckweed) growing on the surface (see figure 8.10).

Examine dishes containing *Lemna* plants set up for you some weeks earlier. Plants in dish A have been growing on distilled water. Plants in dish B have been growing on some pond water obtained from a pond where duckweed normally grows. Apart from the water, the conditions under which they have been grown were kept as similar as possible. At the start of the experiment each dish contained the same number of *Lemna* plants. Which population is growing faster? What would you expect to be a difference between distilled water and pond water as far as ions are concerned? The following experiments may test your suggestion:

a Warm 2 cm^3 of distilled water and 2 cm^3 of pond water in two separate watch glasses over boiling water until all the water has evaporated. Compare the results.

b Take 2 cm^3 of each water as before and add a few drops of dilute nitric acid and one or two drops of silver nitrate solution to each. A white cloudiness indicates the presence of chloride ions.

c Take 2 cm^3 of each water and add a few drops of dilute hydrochloric acid followed by two or three drops of barium chloride solution. A white cloudiness indicates the presence of sulphate ions.

A full analysis of the ions in pond water is impossible in the time available but examine the data in figure 8.11.

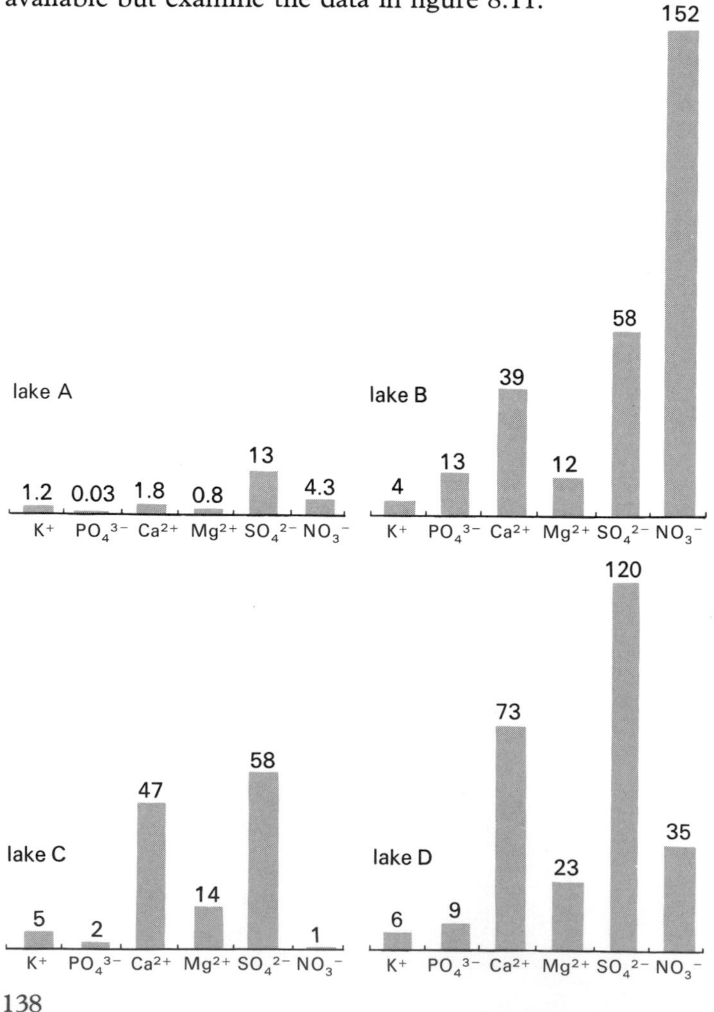

Figure 8.11
Ions in four different lakes
All figures are in mg l^{-1}

These simple experiments and data should convince you that one difference at least between distilled water and pond water is that the latter contains ions. We have also discovered that *Lemna* grows faster in pond water, which contains ions, than in distilled water, which does not. Study the photographs (figure 8.12). These groups of plants have been grown in a series of solutions from which certain ions have been deliberately excluded except for one which contains at least all of the ions mentioned.

Figure 8.12
Plant growth experiments. The jars are labelled with the excluded element. The jar labelled 'complete' contains at least all the elements mentioned

Does a simple pattern emerge? How would you explain it? Why could you doubt the evidence presented in the photographs (which are real enough and have not been faked)? How could you use *Lemna* to test your doubts?

▷ Investigation 8.20 Using fertiliser

▶Predict what would happen if you added a substance such as ammonium sulphate to the soil in which plants were growing.◀ Devise a way of testing your prediction.

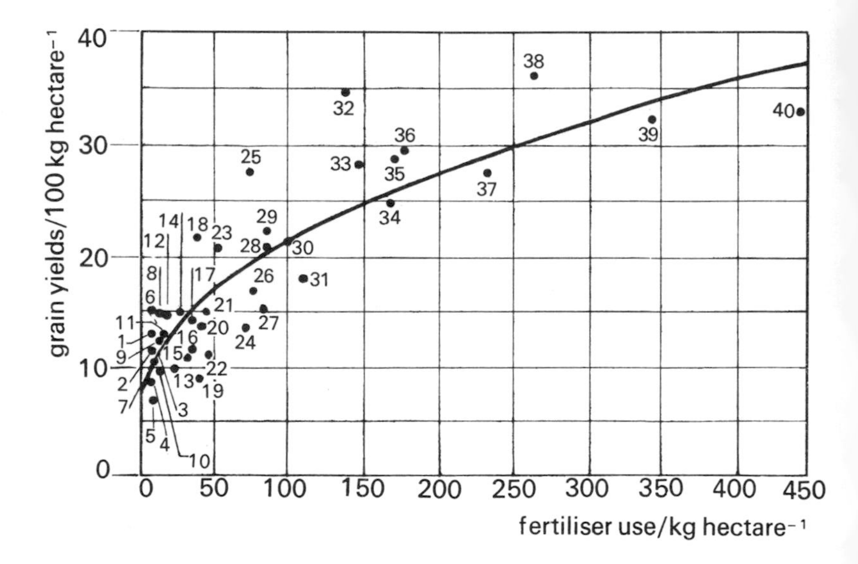

Figure 8.13
Use of fertilisers and yield of grain

Study the graph (figure 8.13). What pattern does it clearly indicate?

Figure 8.14
The plants on the left had molybdenum added to the soil. What conclusion can you draw?

Figure 8.15
Applying fertiliser to the land

Read the following extract:

Among the new techniques which the experts have to offer, one of the most important is the use of fertilisers. It is worth noting that the best-fed countries in the world are also the ones which use the greatest amount of fertiliser. A leading American agricultural scientist, Dr Robert White-Stevens, has estimated that if 10 dollars' worth of fertilisers and 5 dollars' worth of pesticides were used on every acre of the world's arable land world food production would be doubled in the next decade. In 1961, under the Freedom From Hunger Campaign, a programme was launched by FAO in cooperation with the world's fertiliser industry to test and demonstrate the efficacy of the various kinds of fertiliser applied to differing crops and soil conditions. One experiment with paddy rice in India, for example, showed that every ton of nitrogen applied as fertiliser produced more than 13 tons of extra rice. By the end of 1964 more than 25 000 trials and demonstrations had been carried out in 15 countries in Asia, Africa and Latin America, and these are now forming the basis of national programmes. World consumption of fertilisers in 1965–66 was 48.5 million metric tons, an increase of 11.5 per cent on the previous year. In order to persuade peasant farmers to use fertilisers it is usually necessary to bring about a far-reaching change in their attitude to traditional ways of doing things, and so the resulting improvements are often not solely due to the use of fertilisers. It has been estimated that if a fertiliser programme increases output by 100 per cent, half of that increase will be due to the fertilising materials themselves and the other 50 per cent to accompanying changes in farming methods.

Vox Development Despatch no. 2

Discuss the importance of using fertilisers for improving world food production. What other factors are important?

Investigation 8.21 Fluoride ions and dental health

In the first half of this century an interesting pattern was observed in the United States and the United Kingdom. Dentists had discovered that children born and brought up in areas where the local water supply contained small amounts of dissolved calcium fluoride had 50 to 60 per cent less tooth decay than those brought up in areas where the water contained hardly any dissolved calcium fluoride. Although calcium fluoride is almost insoluble in water that which does dissolve forms the ions of calcium and fluoride. It was found that as little as one part per million of fluoride ions in the water was sufficient to produce the pattern.

In 1953 a long-term investigation was started in the cities of Tiel and Culemborg in the Netherlands. The aim was to test the pattern. For ten-and-a-half years fluoride was added to the water supply of Tiel but not to that of Culemborg. Before, during and after this investigation the amount of tooth decay in children of 11 to 15 years of age was measured. What results would you expect? Do the results of the investigation confirm your prediction (figure 8.16). What was the purpose of including Culemborg in the study?

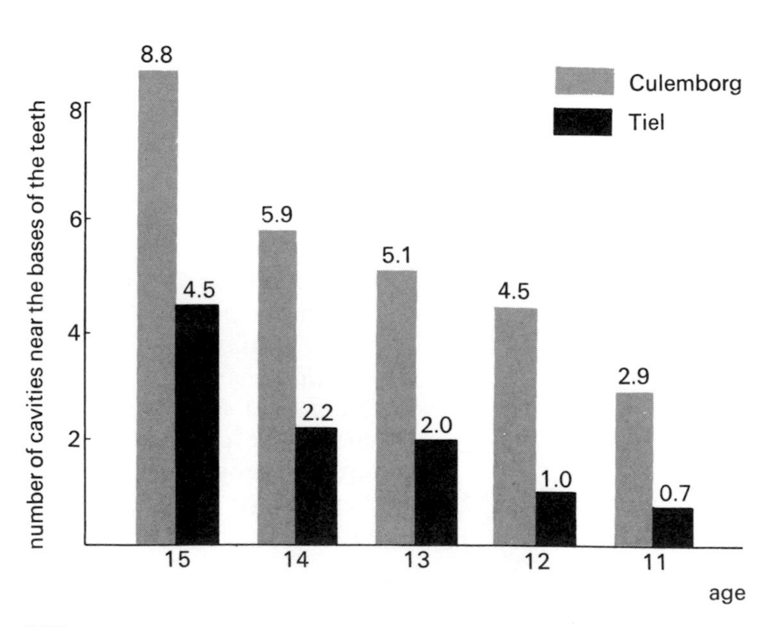

Figure 8.16
Ten-and-a-half years of water fluoridation in Tiel

In Britain three areas have been studied in a similar way—Watford, Kilmarnock and a part of Anglesey. After five years of fluoridation of the water supply the percentage of children who had ten or more decayed teeth at the age of five was reduced from 15 per cent to 2 per cent.

Studies all over the world have had similar results. In 1962, on the basis of these studies, the then Ministry of Health recommended to all Local Authorities that they should add fluoride ions to the water supply if the levels were naturally very small.

In many areas, however, this provoked considerable outcry from the public and local councillors. These are some of the reasons:

1 Concern that fluoridation may cause ill-health.
2 Concern that fluoridation causes flecks of white on the enamel surface of teeth.
3 Concern that fluoridation is 'mass medication' with a loss of personal liberty by compelling people to have fluoride and concern that decisions on such matters should be left to individuals and not taken on their behalf.

The following answers have been given:

1 There is no evidence despite extensive research that fluorides are a cause of ill-health. The following remarks have been made:

> All the established evidence supports the safety of fluoridation although, from the very nature of the problems . . . we can never prove that there are no harmful effects whatever. The evidence does show, however, that if there are any undesirable effects they must be very rare and very mild or they would have been detected by now . . . The unquestionable benefits of fluoridation greatly outweigh the . . . possibility of mild and rare harmful effects.

2 Fluoridation (sufficient to produce a significant reduction in tooth decay) can produce some white flecks in the enamel of teeth. Surveys suggest, however, that in non-fluoride areas problems with the enamel surface of the teeth were far more widespread than in fluoride areas.
3 In any community should there be a small sacrifice of personal liberty for the general good?

What are your views and feelings?

WHAT IS IT? WHAT IS IT?
IT IS A THING OF GREAT MYSTERY, UTTERED BY THE EYEBALLS IN THE SKY
WOOF
UTTER ROT – THERE MUST BE A RATIONAL EXPLANATION, I'LL PROVE IT – SCIENTIFICALLY
FOOLISH BOASTER – HOW CAN YOU DO THAT?
WOOF
I'LL NIP IT!
AIEE! YOU FOOL, YOU FOOL – WHY CAN'T YOU JUST CONTENT YOURSELF WITH TRYING TO BLOW UP THE WORLD?
GRRRRR
B212

Appendix: mathematical notes

Standard form for numbers

In science we often want to refer to, and calculate with, large numbers like 170 000 000 and small numbers like 0.000 000 032. These arise in such varied contexts as estimating numbers and sizes of bacteria, distances to and masses of planets, electrical resistance, numbers and sizes of atoms, molecules or human populations.

There is a convenient standard way of writing such numbers which has at least two advantages. First, it makes the number smaller to look at, and so easier to follow. Secondly it makes calculation easier.

The standard form is to write each number as a number between 1 and 10, multiplied by the appropriate power of 10. This means we write large numbers as follows:

distance to the Moon $\approx 400\,000\,000\,\text{m} = 4 \times 10^8\,\text{m}$
mass of Moon $\approx 75\,000\,000\,000\,000\,000\,000\,000\,\text{kg}$
$= 7.5 \times 10^{22}\,\text{kg}$
typical body resistance $\approx 500\,000\,\Omega = 5 \times 10^5\,\Omega$
speed of light $= 300\,000\,000\,\text{m s}^{-1} = 3.0 \times 10^8\,\text{m s}^{-1}$

To find the power of ten, count how many decimal places to the left the decimal point must be moved to bring it to the correct position i.e. with only one digit to the left of it.

We can apply the same method to small numbers, if we recall that 0.0037 is the same as 3.7/1000 or 3.7×10^{-3}, and similarly for other numbers. For example:

length of a cell $\approx 0.000\,015\,\text{m} = 1.5 \times 10^{-5}\,\text{m}$
wavelength of yellow light $\approx 0.000\,000\,5\,\text{m} = 5.0 \times 10^{-7}\,\text{m}$
estimated length of a molecule $\approx 0.000\,000\,01\,\text{m} = 1.0 \times 10^{-8}\,\text{m}$
mass of an oxygen atom
$\approx 0.000\,000\,000\,000\,000\,000\,000\,000\,027\,\text{kg}$
$= 2.7 \times 10^{-26}\,\text{kg}$

In this case the power of ten is found by noting how many places to the *right* the decimal point must be shifted, and adding a negative sign.

An example of a calculation using standard form is as follows. A rectangular school field is 373 m by 821 m. An estimate is needed of the total number of daisy plants in the field. The average density of daisy plants has been found (from a fairly large sample) to be $15.2\,\text{m}^{-2}$.

$$
\begin{aligned}
\text{total area} &= 373\,\text{m} \times 8.21\,\text{m} \\
&= 3.73 \times 10^{2} \times 8.21 \times 10^{2}\,\text{m}^{2} \\
&= (3.73 \times 8.21) \times 10^{(2+2)}\,\text{m}^{2} \\
&\approx 30.6 \times 10^{4}\,\text{m}^{2} \\
&= 3.06 \times 10^{5}\,\text{m}^{2} \\
\therefore \text{total plants} &\approx 3.06 \times 10^{5} \times 15.2 \\
&= 3.06 \times 10^{5} \times 1.52 \times 10^{1} \\
&= (3.06 \times 1.52) \times 10^{(5+1)} \\
&\approx 4.65 \times 10^{6}
\end{aligned}
$$

So the estimated number of plants is between 4 and 5 million. (Why would it not be sensible to be more precise?)

A slightly more complex example concerns the concentration of a pollutant which has come into a reservoir. If 12.5 kg of mercury has become evenly dispersed through a reservoir 520 m long and 440 m wide and 15.4 m deep, what is the concentration of mercury in the water?

$$
\begin{aligned}
\text{volume of water} &= 520 \times 440 \times 15.4\,\text{m}^{3} \\
&= (5.20 \times 4.40 \times 1.54) \times (10^{2} \times 10^{2} \times 10^{1})\,\text{m}^{3} \\
&\approx (3.52 \times 10^{1}) \times 10^{5}\,\text{m}^{3} \\
&= 3.52 \times 10^{6}\,\text{m}^{3}
\end{aligned}
$$

So concentration of mercury in this water

$$
\begin{aligned}
&\approx 12.5 - (3.52 \times 10^{6})\,\text{kg}\,\text{m}^{-3} \\
&= (1.25 - 3.52) \times (10^{1} - 10^{6})\,\text{kg}\,\text{m}^{-3} \\
&= (1.25 - 3.52) \times 10^{-5}\,\text{kg}\,\text{m}^{-3} \\
&\approx 0.355 \times 10^{-5}\,\text{kg}\,\text{m}^{-3} \\
&= 3.55 \times 10^{-6}\,\text{kg}\,\text{m}^{-3}
\end{aligned}
$$

The density of water is $10^{3}\,\text{kg}\,\text{m}^{-3}$, so the concentration could equally be expressed as

$$
\begin{aligned}
&3.55 \times 10^{-6}\,\text{kg}\,(10^{3}\,\text{kg})^{-1} \\
&= 3.55 \times 10^{-6} \times 10^{-3}\,\text{kg}\,\text{kg}^{-1} \\
&= 3.55 \times 10^{-9}
\end{aligned}
$$

This has no units, as it is just a fraction (the fraction of the liquid which is mercury). It would usually be written as 3.55×10^{-3} parts per million, since

$$3.55 \times 10^{-6} = (3.55 \times 10^{-3}) \times 10^{-6}$$

Here are some simple problems on this way of writing numbers. Write the following in standard form, in each case using the fundamental unit (kg, m, s, W, etc.).

1 3 720 kg
2 0.0031 s
3 0.32 mm
4 1 500 km
5 13.2×10^{-24} g
6 53.2 million
7 443 kN
8 47.3 m s^{-1}
9 95.3 MHz
10 0.47 nm
11 1.42 microgrammes
12 11.4 μm
13 3 200 MW
14 1 371 462 175 s^{-1}
15 0.000 000 000 000 003 g
16 46 720 000 000 000 000 km
17 3.71 ns
18 105 321 N m^{-2}
19 13 600 kg m^{-3}
20 3 900 million years

The advantage in calculation arises from the fact that the powers of ten can be dealt with by adding or subtracting the indices. For example:

$$10^7 \times 10^2 = 10^{(7+2)} = 10^9$$
$$10^{24} \div 10^{11} = 10^{24} \times 10^{-11} = 10^{(24-11)} = 10^{13}$$
$$10^9 \times 10^{-15} = 10^{(9-15)} = 10^{-6}$$
$$10^{-6} \times 10^4 = 10^{(-6+4)} = 10^{-2}$$
$$10^{-23} \div 10^{-7} = 10^{-23} \times 10^7 = 10^{(-23+7)} = 10^{-16}$$

The decimal part of the numbers is dealt with in the usual way, for example by means of a slide rule.

Here are some simple problems to enable you to get used to this system.

1 $10^8 \times 10^5$
2 $10^8 \times 10^{-5}$
3 $10^{-8} \times 10^5$
4 $10^{-8} \times 10^{-5}$
5 $10^{11} \times 10^{-17}$
6 $10^2 \times 10^{-2}$
7 $10^{-3} \times 10^{-5}$
8 $10^{-6} \times 10^{-5}$
9 $10^{-5} \times 10^{-6}$
10 10×10^{-3}
11 $10^7 \div 10^4$
12 $10^7 \div 10^{-4}$
13 $10^{-7} \div 10^4$
14 $10^{-7} \div 10^{-4}$
15 $10^7 \div 10^{-10}$
16 $10^7 \div 10^{-7}$
17 $10^{-7} \div 10^7$
18 $10^{-7} \div 10^{-7}$
19 $10^{-7} \div 10^{-10}$
20 $10 \div 10^{-3}$

21 $(1.2 \times 10^5) \times (8.0 \times 10^2)$
22 $(1.4 \times 10^{-2}) \times (1.1 \times 10^4)$
23 $(1.7 \times 10^{-7}) \times (1.2 \times 10^{-3})$
24 $(5.0 \times 10^6) \times (5.0 \times 10^7)$
25 $(4.1 \times 10^{-1}) \times (4.7 \times 10^{11})$
26 $(1.2 \times 10^5) \div (8.0 \times 10^2)$

27 $(1.8 \times 10^{-2}) \div (2.0 \times 10^{3})$
28 $(1.5 \times 10^{-7}) \div (5.0 \times 10^{-2})$
29 $(3.7 \times 10^{2}) \div (5.2 \times 10^{-6})$
30 $(8.9 \times 10^{27}) \div (9.1 \times 10^{-3})$

▷A third example is even more complex. Suppose we needed to know the approximate volume of a molecule, and we had estimated its length as 1.0×10^{-8} m, and its breadth and height as both 4.0×10^{-10} m. Then, assuming it can be regarded as a rectangular box (the estimates are clearly very approximate):

$$\begin{aligned}\text{volume} &\approx (1.0 \times 10^{-8}\,\text{m}) \times (4.0 \times 10^{-10}\,\text{m}) \times (4.0 \times 10^{-10}\,\text{m})\\ &= (1.0 \times 4.0 \times 4.0) \times (10^{-8} \times 10^{-10} \times 10^{-10})\,\text{m}^3\\ &= 16.0 \times 10^{-28}\,\text{m}^3\\ &= 1.6 \times 10^{-27}\,\text{m}^3\end{aligned}$$

Similarly we can work out the size of an atom of silver from the knowledge that the mass of 6×10^{23} atoms of silver is 1.08×10^{-1} kg and the density of silver is $1.05 \times 10^{4}\,\text{kg}\,\text{m}^{-3}$.

0.108 kg of silver contains 6×10^{23} atoms

$$\begin{aligned}\therefore 1.05 \times 10^{4}\text{ kg of silver contains } 6 \times 10^{23} \times \frac{1.05 \times 10^{4}}{1.08 \times 10^{-1}}\text{ atoms}\\ \approx 6 \times 10^{23} \times 10^{5}\text{ atoms}\\ = 6 \times 10^{28}\text{ atoms}\end{aligned}$$

I.e. $1\,\text{m}^3$ of silver contains 6×10^{28} atoms.

Now a cube of 1 m side would have n atoms per side, where

$$\begin{aligned}n \times n \times n &= 6 \times 10^{28}\\ &= 6 \times 10 \times 10^{27}\\ &= 60 \times (10^{9})^{3}\\ &\approx 4^{3} \times (10^{9})^{3}\end{aligned}$$

So $n \approx 4 \times 10^{9}$.

This means there are 4×10^{9} atoms in 1 m, so the diameter of each atom

$$\begin{aligned}&= \frac{1}{4 \times 10^{9}}\,\text{m}\\ &= \tfrac{1}{4} \times 10^{-9}\,\text{m}\\ &= 0.25 \times 10^{-9}\,\text{m}\\ &= 2.5 \times 10^{-10}\,\text{m}\end{aligned}$$

You might like to try the following:

1 If 12 million homes in Britain use a total of 2.4×10^{15} J of energy per day, what is the average energy per hour used per home throughout the day?

2 What is the average density of the Moon, given that its volume is $2 \times 10^{19}\,m^3$ and its mass is 7.5×10^{22} kg?

3 How many red blood cells would there be in your body if the diameter of a red blood cell is about 8×10^{-6} m and its thickness about 2×10^{-6} m? Your body contains about $3 \times 10^{-3}\,m^3$ (3 litres) of blood, and about 0.1 of it consists of red cells.

The use of the slide rule

Most modern mathematics courses include the slide rule at some stage, usually giving the theory behind it as well as the way to use it. Here the slide rule is just treated as a tool for calculating—if you like you can regard it as a very simple sort of computer. You can use it without understanding the theory, but understanding the theory will help you to use it better.

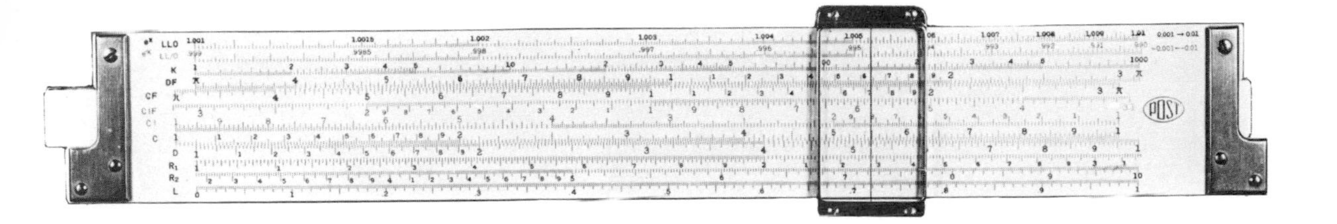

A slide rule consists of a stock, or body, and a slide, which is made to fit the stock rather precisely. In addition there is a cursor, which is a transparent plate which can be moved along the stock, covering both stock and slide. The cursor has a fine line engraved on its under surface.

On both stock and slide various scales are engraved. Expensive slide rules have up to twenty scales of high precision. Cheaper ones have fewer scales, and usually they are not so precise. It is a mistake to start with an over-complex set of scales, because a few standard ones are sufficient for most school calculations. Perfectly adequate slide rules can be obtained for less than £1.

Most ordinary slide rules have scales which are either 125 mm or 250 mm long. The most important scales are called the C-scale and the D-scale. These are used for multiplication and division. The C-scale is on the slide, and the D-scale on the lower part of the stock. Otherwise they are identical. If you look at a slide rule you will see that the numbers are not equally spaced, being more crowded at the right hand side (higher numbers).

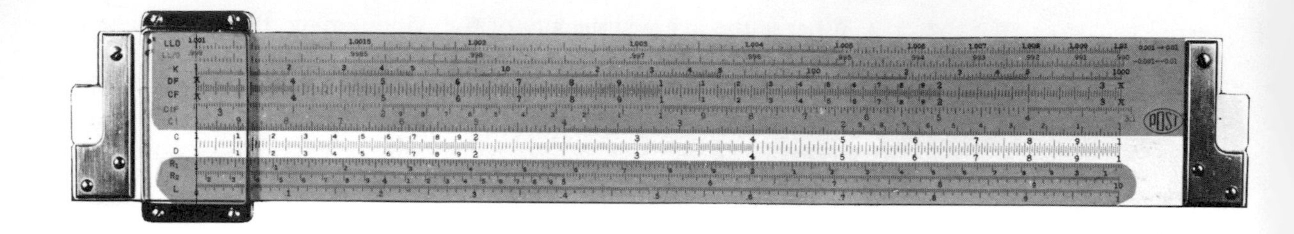

If you count downwards from 9 to 2 you will notice that the sub-divisions vary at different parts of the scale. Below 2 the numbers 9 down to 1 are usually engraved again, in smaller figures, representing 1.9 down to 1.1. You must make sure you do not mistake these for the larger numbers representing 9 down to 2. At the extreme left is the number 1, and at the extreme right the number 10. On some rules instead of 10 you will find another 1. This is because although the scale runs from 1 to 10, it can just as well be thought of as running from 0.1 to 1, 10 to 100, 100 to 1 000 or any similar range. Only the digits matter, the position of the decimal point is unimportant. (We shall come back to this later.)

Part a Multiplying two numbers using the C and D scales

Suppose you want to find the area of a field which is 215 m by 142 m. How you go about this depends on how accurate an answer you need.

To work out how many cattle you could graze on it, probably all that would be necessary is the following:

$$\begin{aligned} \text{area} &= 215 \times 142\,\text{m}^2 \\ &\approx 200 \times 150\,\text{m}^2 \\ &= 30\,000\,\text{m}^2 \end{aligned}$$

Notice that the rounding off increases one figure and decreases the other, reducing the overall error which is introduced.

If you wanted to compare the yield of a crop (say wheat) from this field with that from another field, a more accurate calculation would be needed. It might be sufficient to say

$$\begin{aligned} \text{area} &= 215 \times 142\,\text{m}^2 \\ &\approx 210 \times 150\,\text{m}^2 \\ &= 31\,500\,\text{m}^2 \text{ (by long multiplication)} \end{aligned}$$

or

$$\begin{aligned} \text{area} &= 215 \times 142\,\text{m}^2 \\ &\approx 220 \times 140 \\ &= 30\,800\,\text{m}^2 \end{aligned}$$

But using a slide rule it would be quicker to say

$$\text{area} = 215 \times 142\,\text{m}^2$$
$$\approx 30\,500\,\text{m}^2$$

How is this done? The picture opposite shows a slide rule on which only the C-scale and the D-scale have not been shaded.

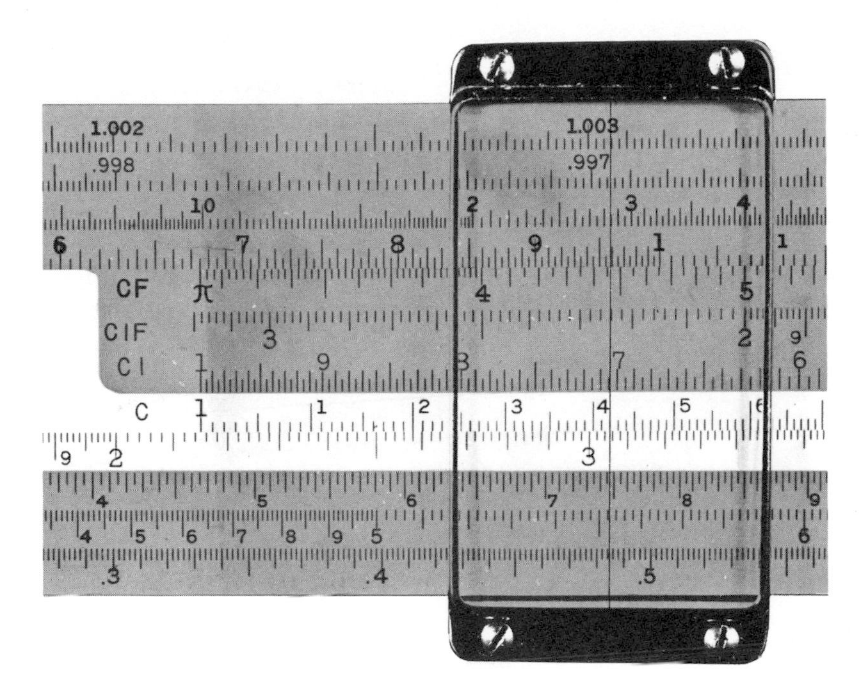

The operations are as follows:

1. Move the cursor so that the cursor line is over 2.15 on the D-scale.
2. Move the slide so that 1.00 on the C-scale is beneath the cursor line. (This gives a position of the slide in which every number on the C-scale is opposite *2.15 times that number* on the D-scale.)
3. Move the cursor so that the cursor line is over 1.42 on the C-scale. (This is opposite 2.15×1.42 on the D-scale.)
4. Read off the number from the D-scale beneath the cursor line. (Between 3.05 and 3.06, slightly nearer the former.)
5. Put the decimal point in the correct place for the original calculation. (Area $\approx 30\,500\,\text{m}^2$.)

The last step is in practice often done first, by doing a rough calculation like the first one above ($200 \times 150\,\text{m}^2 = 30\,000\,\text{m}^2$). The slide rule gives the *digits* in the answer (to three figures in most cases) but can never give the position of the decimal point.

Put another way, the slide rule will do calculations like

$2.15 \times 1.42 \approx 3.05$

but you have to put in the powers of ten yourself. For example calculations like 215×142 would be done as follows:

$$
\begin{aligned}
215 \times 142 &= (2.15 \times 10^2) \times (1.42 \times 10^2) \\
&= 2.15 \times 1.42 \times 10^4 \\
&\approx 3.05 \times 10^4
\end{aligned}
$$

This is very easy once you have got used to the idea.

Part b Multiplying two numbers using the C and D scales (off-scale answer)

If the field above had been a different shape and size, say 215 m × 742 m it would not be possible to work just as above. The picture shows why. The 7.42 on the C-scale is right off the D-scale.

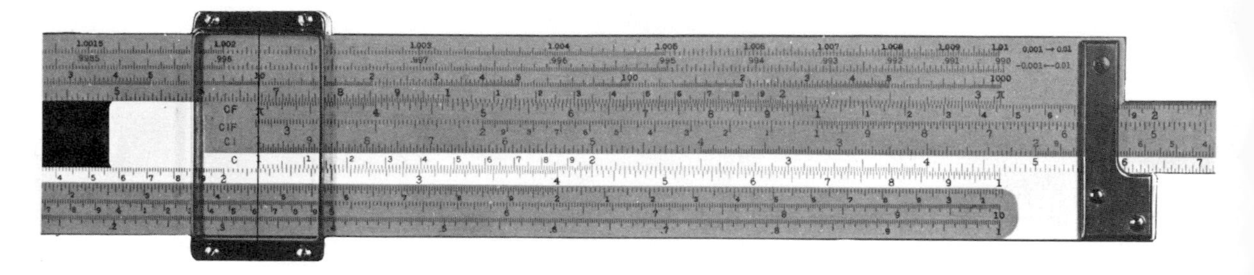

This will happen whenever the product is greater than 10.0. The way to deal with this is just to use the 10.0 (or right-hand 1.00) on the C-scale instead of the left-hand 1.00. This is shown below.

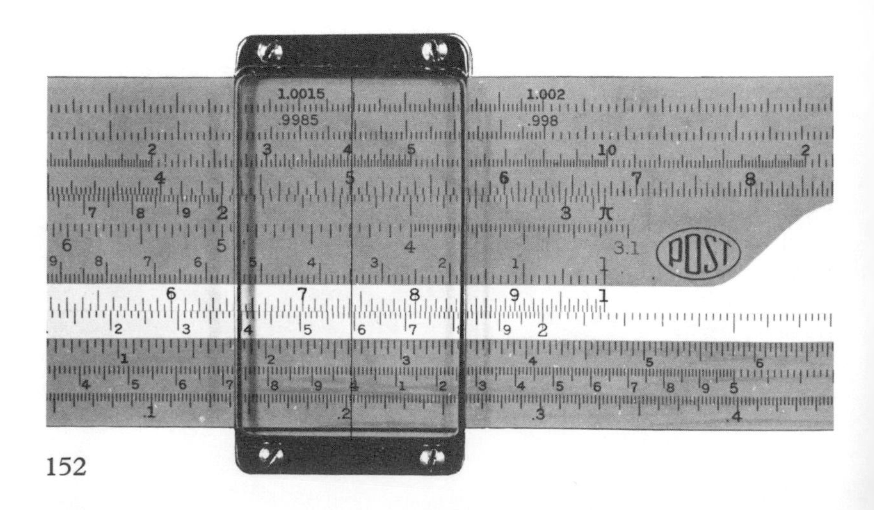

1 Put the cursor line over 2.15 on the D-scale.
2 Put the 10.0 on the C-scale beneath the cursor line.
3 Put the cursor line over 7.42 on the C-scale.
4 Read off the answer from the D-scale beneath the cursor line.
The 7.42 on the C-scale is now found to be opposite 1.59 on the D-scale, so we have

$$\begin{aligned} 215 \times 742 &= 2.15 \times 7.42 \times 10^4 \\ &\approx 15.9 \times 10^4 \\ &= 1.59 \times 10^5 \end{aligned}$$

The area of this field is about $1.59 \times 10^5 \text{ m}^2$.
(The rough calculation would be

$$\begin{aligned} 215 \times 742 &\approx 200 \times 750 \\ &= 150\,000.) \end{aligned}$$

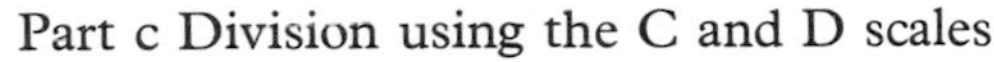

Part c Division using the C and D scales

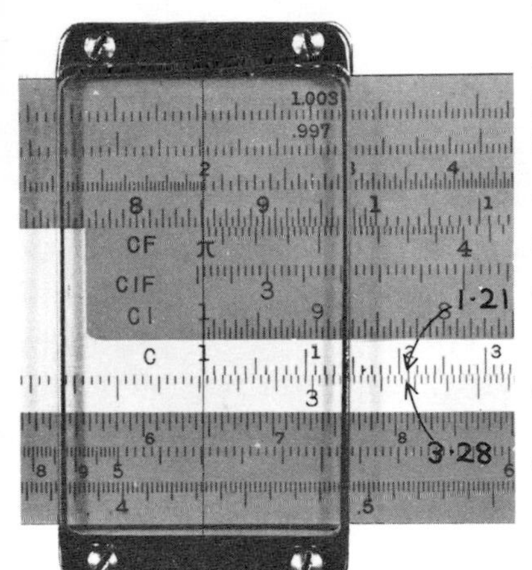

In contrast to ordinary calculations, division is easier than multiplication on a slide rule. You might have the following data from an experiment.

volume of aluminium cylinder: 121 cm^3
mass of cylinder: 328 g

To find the density of aluminium you say

$$\text{density} \approx \frac{328}{121} \text{ g cm}^{-3}$$

The method of doing this calculation on the slide rule is
1 Put the cursor line opposite 3.28 on the D-scale.
2 Put 1.21 on the C-scale opposite the cursor line.
3 Move the cursor line to 1.00 on the C-scale and read off the answer from the D-scale (2.71).
This is illustrated on the left.
The rough calculation gives 300/100 = 3.

$$\begin{aligned} \text{So density} &\approx \frac{328}{121} \text{ g cm}^{-3} \\ &\approx 2.71 \text{ g cm}^{-3} \\ &= 2\,710 \text{ kg m}^{-3} \end{aligned}$$

(since $1 \text{ g cm}^{-3} = 1\,000 \text{ kg m}^{-3}$).

Part d Division using the C and D scales (off-scale answer)

If a smaller piece of aluminium had been used the results might have been

volume of aluminium cylinder: 82 cm^3
mass of cylinder: 225 g

The operations are just as before except that the answer is read off opposite 10.0 on the C-scale, as shown below.

Notice that in division it is never necessary to re-position the slide. All you do is read the answer opposite 1.0 or 10.0, whichever is on the scale. This is why division is easier than multiplication.

Part e Multiplication using CI and D scales

On many slide rules the slide has a CI-scale, which is a C-scale engraved in reverse. This scale represents the *reciprocal* (or *inverse*) of the C-scale: CI stands for C-Inverse. With this scale you can multiply using the process for division, since

$215 \times 142 = 215 \div (\frac{1}{142})$

This is illustrated opposite. Note that the answer is opposite the 1.0 at the *right-hand* end of the CI-scale.

The answer is always on the scale because if 1.0 is off scale the answer is found opposite 10.0, as with division. This method is therefore simpler than the one given earlier.

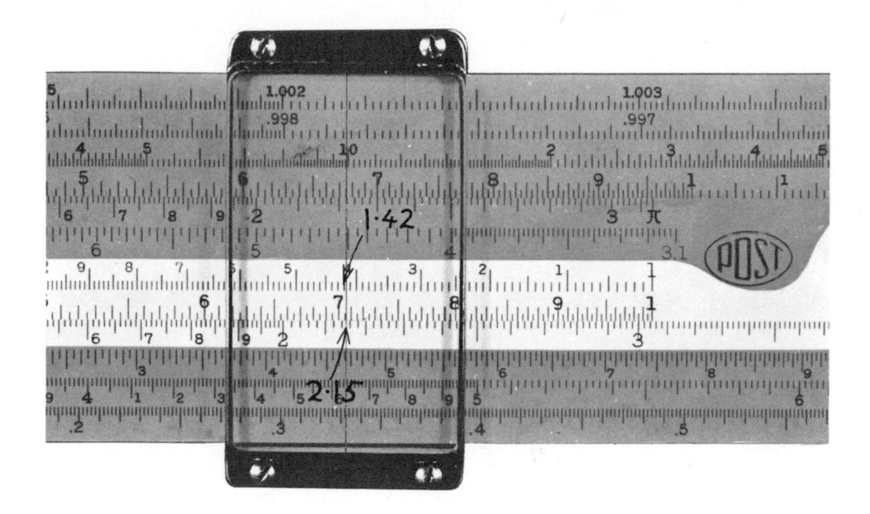

The CI-scale is usually in the centre of the slide, not next to the D-scale. This means that the use of the cursor is essential, whereas in previous work it was merely a useful extra.

You might wonder whether the CI-scale can be used for division. This is possible, but there is no point in doing so, because division using the C- and D-scales is easier than it would be using the CI- and D-scales.

Part f More complex calculations

If you need to work out

$$\frac{73.7 \times 1\,900 \times 0.316}{1.82 \times 127}$$

this can be done in a series of related steps, without finding any intermediate answers. The simplest and easiest approach is shown below.

1 Find approximate answer:

$$\frac{70 \times 2\,000 \times 0.3}{2 \times 100} = 210$$

2 $7.37 \div 1.82$, place cursor line over result ($= R_1$) (4.05)
3 $R_1 \times 1.9 \rightarrow R_2$ (7.7)
4 $R_2 \div 1.27 \rightarrow R_3$ (6.06)
5 $R_3 \times 3.16 \rightarrow R_4$ (1.91)
6 Insert decimal point: Answer = 191. None of the intermediate answers would be recorded, or in fact read off. The cursor line acts as a temporary memory until the next operation is done.

Index

Bold figures indicate illustrations.